BRITISH URBAN POLICY
An Evaluation of the Urban Development Corporations

BRITISH URBAN POLICY
An Evaluation of the Urban Development Corporations

edited by

ROB IMRIE AND HUW THOMAS

SECOND EDITION

SAGE Publications
London • Thousand Oaks • New Delhi

First published 1999

 A SAGE Publications Ltd
6 Bonhill Street
London EC2A 4PU

SAGE Publications Inc.
2455 Teller Road
Thousand Oaks, California 91320

SAGE Publications India Pvt Ltd
32, M-Block Market
Greater Kailash-I
New Delhi 110 048

British Library Cataloguing in Publication Data
A catalogue record for this book is available from the British Library

ISBN 1–7619–6225–5
ISBN 1–7619–6226–3 (pbk)

Library of Congress catalog card number available

Typeset by Anneset, Weston-super-Mare, Somerset
Printed and bound in Great Britain by Cromwell Press, Wiltshire

Contents

Preface

The publisher's invitation to submit a proposal for a second edition of this book illustrates the continuing interest in one of the most important urban policy initiatives in Britain over the last twenty years, an initiative which has courted controversy and political disputation throughout its life. The Urban Development Corporations (UDCs), introduced into the political bear-pits of Merseyside and east London in the early 1980s with some predictable hostility, have managed to garner headlines in the late 1990s as they exit amid accusations of incompetence and lack of cooperative working.

The UDCs were introduced as Thatcherite flagships, espousing a disdain for the efficiency and effectiveness of local government and a belief that the only practical approach to addressing urban problems was through creating conditions in which markets could function. In the case of UDCs, the task was seen, initially, as one of boasting property markets which would draw in private investment to revitalise the designated urban development areas (UDAs). The costs involved in attracting and then underpinning, investor confidence in fragile property markets was to be enormous, as Chapter 1, and the case studies in this book, illustrate.

Moreover, the distribution of short-term benefits of that expenditure was so evidently skewed away from the generally poor populations of UDAs, and their surrounding areas, that the London Docklands Development Corporation (LDDC), in particular, drew adverse comment from a House of Commons committee (see Brownill, Chapter 2). The LDDC responded to the criticism, as did other UDCs (designated in so-called 'generations' from the mid-1980s onwards). While the basic UDC approach – property-led, market-sensitive – has remained in place, there have also been changes over time, and across space, with individual UDCs responding to changes in the property market (and economic conditions more generally) and varying local political circumstances.

A number of changes can be discerned in the second edition of the book. The first chapter has been updated and amended in the light of changes in UDC policy, and the emergence of new understanding about their roles and impacts in the British cities. Four chapters have been dropped from the first edition, not because of any inadequacies on their part but because we wanted to draw in the stories of other UDCs which had not been covered in the original book. Thus, the second edition contains new chapters on the experiences of the UDCs in Teesside and Central Manchester while Allan Cochrane has contributed a new, concluding, piece which seeks to assess the legacies of the UDCs. All of the chapters from the previous edition which have been retained, have been updated and, in most instances, restructured to take into account the changing circumstances of the respective UDCs. The second edition then, provides a series of fresh perspectives on the UDCs, although the underlying structure of the first edition has been more or less retained.

In producing this revised edition we would like to thank the individual contributors who were enthusiastic and willing and produced their chapters to tight deadlines. We would also like to thank the Nuffield Foundation who provided funding for a workshop held in September 1992 which provided the original inspiration for the first edition of the book. This edition could not have been produced without the support of a number of individuals and we would like to acknowledge the cartographic skills of Justin Jacyno, in the Department of Geography at Royal Holloway, and Janice Coles in the Department of City and Regional Planning, Cardiff University. Alison Simmons, in Cardiff, had the onerous task of producing the final manuscript to a very tight deadline – a task often made more, rather than less, frustrating by the disks provided by contributors! The book could not have been produced so smoothly without her efforts. Andrew Edwards, her colleague, helped her overcome technical difficulties in retrieving reluctant texts. We are particularly grateful to Jeanette Graham, and the editorial team, for their endeavors and hard work in facilitating the final production of the book.

Rob Imrie, University of London
Huw Thomas, Cardiff University
June, 1998

The Editors

Rob Imrie is Reader in Human Geography at Royal Holloway, University of London. He has published widely in international journals on subjects spanning urban policy to industrial and economic planning, and disablism and planning. He is co-author of one book (1992, Transforming Buyer-Supplier Relations, Macmillan, London), and author of another (1996, Disability and the City: International Perspectives, Paul Chapman Publishing, London, St. Martin's Press, New York). At present he is directing an ESRC-funded project on *property markets and disabled access in Sweden and the UK*.

Huw Thomas is a Senior Lecturer at the Department of City and Regional Planning, Cardiff University. He has published widely in international journals in planning, geography and urban studies. Recently, he co-edited *Urban Planning and the British New Right* (1998, Routledge, London). His current research interests include the racialisation of planning and urban policy (on which he is undertaking EU funded research) and nationality and planning.

Notes on Contributors

Michael Bradford is Professor of Geography and a member of the Centre of Urban Policy Studies at the University of Manchester. His research interests are in urban and social policy analysis and evaluation, educational restructuring, and the geography of children. He recently completed an evaluation of three UDCs and is currently working on an update of the Index of Deprivation, along with research projects on the business of children's play, the abuse of children by strangers, and a 1990s race discrimination suit in the USA.

Sue Brownill is Principal Lecturer in the School of Planning, Oxford Brookes University. She previously worked for the Docklands Forum, a community planning organisation in London Docklands. In addition to the London Docklands her research interests include the governance of urban policy, public participation in planning and urban policy, and race, gender and regeneration.

David Byrne is Senior Lecturer in Social Policy at the University of Durham. His interest in urban policy and its social consequences originated in his period as Research Director of the North Tyneside CDP in North Shields. Current research interests include collaborative work with colleagues in Poland on the restructuring of Upper Silesia, one of Europe's most important industrial districts, and work on the application of ideas derived from the fields of Chaos and Complexity to the understanding of social change.

Allan Cochrane is Professor of Public Policy in the Faculty of Social Sciences at the Open University. He has written widely in the fields of urban policy and local government. He is the author (with John Allen and Doreen Massey) of *Re-thinking the Region,* published by the Open University Press in 1998. He is the editor (with John Clarke and Eugene McLaughlin) of *Managing Social Policy* published by Sage in 1994 and (with James Anderson and Chris Brook) of *A Global World?* published by Oxford University Press in 1995.

Bob Colenutt is Head of the Urban Regeneration Division at the London Borough of Haringey. He worked for community action groups on the South Bank and London Docklands between 1972-84 and in the Docklands Team of the GLC between 1984-86. He was Head of the Docklands Consultative Committee Support Unit between 1986-93 and of the Thames Gateway Unit at LB Barking and Dagenham, between 1993-95. He has written widely on community led planning and regeneration.

Gordon Dabinett is the CRESR Reader in Urban & Regional Policy at the School of Urban & Regional Studies, Sheffield Hallam University, and is the current Chair of the Regional Studies Association. Previously he worked in economic development for Sheffield City Council. His interests focus on aspects of local economic development and technological change, and he has been involved in various urban and regional policy evaluations.

Iain Deas is a Lecturer in the Department of Planning & Landscape at the University of Manchester. Recent research has included Government-commissioned projects to evaluate a variety of urban regeneration initiatives, including work on UDCs. He is currently working on a project funded under the ESRC's Cities programme to explore the relationship between competitiveness and social cohesion in Liverpool and Manchester.

Marty Lawrence is a research officer at Newcastle City Council and is undertaking a PhD at Oxford Brookes University on the health impacts of housing policies. His current interests include urban policy, deprivation and health.

Andrew May is an Associate Lecturer at the University of the West of England, Bristol, Faculty of the Built Environment, Housing and Urban Studies School. He is a member of Bristol City Council. From 1984 to 1997 he was chair of Bristol Planning and Development Committee.

Richard Meegan is a Senior Lecturer in the Department of Geography in the University of Liverpool. He joined the Department after working at the Greater London Council and the Centre for Environmental Studies, two institutions deemed worthy of closure by Conservative central government administrations, and the independent non-profit research centre CES Ltd, London. His interests are in urban and regional studies and the geography of social and economic restructuring with a more recent emphasis on issues of community-based economic development and social exclusion.

Nick Oatley is a Senior Lecturer at the University of the West of England, Bristol, Faculty of the Built Environment, Housing and Urban Studies School. He is currently involved in the ESRC research programme on 'Cities: Competition and Cohesion' looking at governance practices in regeneration activities and contemporary forms of urbanisation and the emergence of new urban spaces. He recently published an edited book on Cities, *Economic Competition and Urban Policy* (Paul Chapman, London).

Jamie Peck is Professor of Geography and Director of the International Centre for Labour Studies at the University of Manchester. His research interests are in urban political economy, labour market restructuring and the politics of economic regulation. He is currently working on a series of projects concerned with welfare reform, welfare-to-work policies and the emergence of workfare strategies in Britain, Canada and the United States.

Peter Ramsden is head of Enterprise plc's Brussels Office where he has responsibility for European business development for the company. His main areas of focus are local and regional development, innovation, information society and financial engineering. Between 1994 and 1996 he worked as a national expert with DDG XVI of the European Commission. Previously he was a senior lecturer in Urban Policy and member of CRESR at Sheffield Hallam University.

Fred Robinson is a Research Lecturer, Department of Sociology and Social Policy, University of Durham. His main research interests centre on the social impacts of economic change and processes, politics and practice in urban regeneration, with particular reference to North East England.

Keith Shaw is a principal lecturer in Government and Politics at the University of Northumbria and is a member of the management board of the University's Sustainable Cities Research Institute. His research includes urban policy evelution, local economic development, regional governance, community involvement and Local Agenda 21 initiatives.

Adam Tickell is Professor of Geography and ESRC Research Fellow at the University of Southampton. His research interests are in the political and cultural economy of the financial sector, urban restructuring, and socio-economic regulation. He is currently working on changes in the dominate cultural mores and the political economy of concentration in the financial sector in Britain, Canada and the United States.

Kevin Ward is a Research Associate in the International Centre for Labour Studies and member of the School of Geography at the University of Manchester. His research interests are in urban political economy, the politics of city regeneration and the management of work. He is currently working on a project concerned with the relationship between employment change in large organisations and local labour markets in the North West of England.

enabled successive Conservative governments in their efforts to redefine the content and direction of urban policy. The idea of a complex interplay of economic, social and environmental factors, each requiring an appropriate response, exemplified by the aspirations, if not reality, of the original inner city UP authorities and partnerships, was also being replaced by a focus on physical transformation, a property-led approach to regeneration. Policy instruments, like City Grants and the UDCs, were exemplars of this approach, investment in buildings and infrastructure, based on the unsubstantiated premise that their supply would (inevitably) generate new jobs and wealth. The assumptions of this approach, in turn, made it easier to involve the private sector, and certainly private sector property development in inner cities has been easier to encourage (with suitable subsidies) than, say, business involvement in city technology colleges.

The involvement of business interests in urban policy was also facilitated by local government responses to global economic changes and their uneven impact on particular areas. Harvey (1987), Mayer (1989) and others have noted the increase in urban entrepreneurialism, as local authorities and public agencies responded to rapid shifts in international patterns of investment. Places need to compete for mobile investment, and this means, inter alia, creating the 'right business climate'. This has many aspects, but of crucial significance is that public agencies, including local government, be seen to be responsive to, and closely working with, businesses in the community. Thus, promotional literature and place marketing was increasingly seen as important. In some parts of Britain, local businesses have organised their own strategies for creating the right conditions for growth. In general, the 1980s, with its place marketing, saw closer relationships developing between local government and business interests, irrespective of central government pressure to increase private sector influence (see Bailey, 1995). The UDCs, with their private-sector-dominated Boards, represented a logical continuation of this trend.

The content and direction of the property-based approach has attracted much comment and criticism (Imrie and Thomas, 1992, 1993a; Lawless, 1991; Oatley, 1998; Solesbury, 1990; Turok, 1992 also, see Hansard 1992a, b and c). In a wide-ranging review, Turok (1992) concludes that property development is hardly a panacea for economic regeneration and, even in physical terms, appears to be deficient. For instance, the Audit Commission (1989) noted that the rate of reclamation of derelict land during the 1980s failed to keep pace with the growth of new derelict spaces. Similarly, Turok notes that property-based measures ignore some of the crucial dimensions of city revitalisation, such as education and training, investment in basic infrastructure (like transport and communications), and the underlying competitiveness of industry (especially the technical bases of production and the innovative capacity of firms). However, the difficulty here is not necessarily the weaknesses of property development per se, but the relative absence of additional strategies and measures in revitalising cities.

It is also apparent that government emphasis on property-led regeneration was, in part, an attempt to facilitate local economic growth. Statements from the DoE gave formal backing to the idea that physical redevelopment provided a major stimulus in economic restructuring, yet evidence suggests that a property-based approach cannot guarantee any rise in the level of economic activity. As Turok (1992) rightly argues, one of the main problems of property-based approaches in facilitating economic regeneration is likely to be that the other preconditions for growth are absent or weak in the target cities. However, Healey (1992) notes that urban policy was increasingly concerned with propagating property development as the purveyor of new rounds of economic growth, solely on the assumption that the new physical spaces will attract in the 'right' mix of investors. Yet, as the collapse of the London Docklands' developer, Olympia and York, testified, many of these assumptions were built on shifting sands.

Others have also criticised the general property-based approach and its distributional consequences (NAO, 1988, 1993). For instance, the Public Accounts Committee in 1988 called for UDCs to place more emphasis on strategic issues, while others condemned the concept of leverage planning as an ineffectual tool of city revitalisation (CLES, 1990a and b; Imrie and Thomas, 1992; Thomas and Imrie, 1989, 1997). Brownill (1990), for instance, shows how the haste to maximise private investment led to developers receiving substantial public subsidies, while others note that the primacy of the approach, in seeking maximum developers' profit, precipitated 'runaway' developments characterised by the absence of strategic infrastructure, like highways and open space (Edwards, 1997; Imrie and Thomas, 1992; Rowley, 1994). Harding (1991), in a wide-ranging review of regeneration strategies, highlights the fact that property-led strategies in the 1980s appeared to ignore local needs.

Yet, the themes of urban competition, regional autarchy, and private sector influence have, if anything, been strengthened by other policy initiatives in the 1990s, such as City Challenge, City Pride, and, most significantly, the Challenge Fund (see Edwards, 1997). In particular, the process of competitive bidding for funds, which underpinned City Challenge, and which is a staple basis of the Challenge Fund, required local authorities to demonstrate broad-based business and community support and involvement in their renewal strategies. The dominance of the values of the market has been noted by Gosling (1992, p. 24) who described the City Challenge process as:

> a sweepstake dubbed 'Tarzan's tombola' with many officers involved in the bidding complaining that a 'hard sell' presentation is worth more than the quality of the schemes.

City Challenge typified over a decade of urban policy, part of a pastiche which was financed by top-slicing or cutting other UP funds. Indeed, the UP was cut by £15 million and housing investment by a further £45 million, in 1991–92, to help finance City Challenge. However, City Challenge, as others have documented, marked a new phase of policy which, far from seeking to work against the local authorities, provided opportunities for their incorporation and involvement, albeit in ways strictly defined by central government (Oatley, 1995, 1998).

City Challenge, then, seemed to signify the partial abandonment of a property focus to urban regeneration in favour of a broader, more socially inclusive, set of policies. Such shifts were also discernable in UDC strategy and, by the mid-1990s, the development corporations were increasingly involved in partnerships, while social, community, and environmental projects were much more evident than had hitherto been the case. By early 1998 (at the time of writing), the Blair administration has moved the urban policy agenda forward with the possibilities of a democratisation of policy processes, while seeking to develop integrated programmes which interconnect, for example, education and training provision with transport and urban and regional policy initiatives. However, for Edwards (1997), the emergent phase of urban policy owes much to the UDCs' emphasis on a culture of competition, so providing a source of continuity with the past (see Oatley, 1998, for an overview of the new policy regimes).

Urban policy and the urban development corporations: policies and practices

There is a wide range of literature describing the origins, ethos, and objectives underlying the British UDCs (Batley, 1989; CLES, 1990a and b; Imrie and Thomas, 1993b; Lawless, 1989; O'Toole, 1996; Parkinson, 1990; Thomas and Imrie, 1997; Thornley, 1990, 1991). The UDCs became a central institutional mechanism of British urban policy from the early 1980s, and, as Lawless (1991) notes, they were appointed by central government to oversee the physical regeneration of specific localities, primarily by bypassing the traditional deliverers of urban policy, local government. Their original remit was set out by the 1980 Act, with the focus on property-led regeneration made clear:

> to secure the regeneration of its area by bringing land and buildings into effective use, encouraging the development of existing and new industry and commerce, creating an attractive environment and ensuring that housing and social facilities are available to encourage people to live and work in the area. (Section 136)

Armed with a range of land acquisition and planning control powers, the UDCs claimed to have been able to circumvent local government, avoiding controls which, so successive Conservative governments argued, were at the heart of urban decay in the British cities. In this sense, the first Thatcher administration devised UDCs as 'trouble-shooting' organisations, although various commentators have criticised their methods of evading local democracy, remaining unaccountable to local electorates, and creating inequitable forms of urban development (Fainstein *et al.*, 1992; Lawless, 1989; Parkinson and Evans, 1988 and 1989; Thornley, 1991).

The single or dedicated agency approach to regeneration has been widely documented, and became an important feature of the post-war planning of the New Towns through the New Town Development Corporations, agencies which were given extensive powers to acquire and develop land (see Cullingworth and Nadin, 1997; Ward, 1995). As Lawless (1989) notes, such was the success of the New Towns programme that the Town and Country Planning Association (1979) made several calls for the designation of development corporations to solve the problems of the inner cities. Such exhortations obviously appealed to the incoming Conservative government of 1979 who drew connections between independent, centrally appointed development agencies, free from the apparent constraints of local government, and the creation of conducive conditions for the operation of unfettered market forces in securing the reconstruction of the cities.

A range of authors concur that the UDCs were illustrative of the dominant themes of emergent British urban policy in the 1980s, that is, the minimisation of local state agencies in urban regeneration and creating a climate conducive to the private investor (Batley, 1989; Stoker, 1991; Thornley, 1991). This ethos was outlined by a House of Lords Select Committee (1981, p. 7) which contended that the rationale of the UDCs was 'to remove the political uncertainty and restraints of local democracy which . . . represents a significant hindrance to the development process and a deterrent to private investment'. Indeed, there is much evidence to suggest that the UDCs were at the forefront in the restructuring of central-local government relations. In reflecting on the designation of the London Docklands Development Corporation (LDDC), Michael Heseltine commented on the displaced local authorities within the Urban Development Areas:

> we took their powers away from them because they were making such a mess of it. They are the people who got it all wrong. They had advisory committees and even discussion committees, but nothing happened . . . UDCs do things and they are free from the delays of the democratic process. (Hansard, 1987)

For Heseltine, then, development corporations represented a new dawn and, as Table 1.1 and Figure 1.1 indicate, there were 13 UDCs in operation in Britain by early 1993, designated in four phases or generations. The phase one UDCs, comprising the flagship, LDDC, and Merseyside Development Corporation (MDC), were set up in 1981. Five more followed at the beginning of 1987, comprising Teesside (TDC), Tyne and Wear (TWDC), Trafford Park (TPDC), the Black Country (BCDC), and Cardiff Bay (CBDC). These were followed in 1988 with the announcement of the so-called 'mini-UDCs' for Central Manchester (CMDC), Bristol (BDC), Sheffield (SDC) and Leeds (LDC) (with the extension of the Black Country UDC into Wolverhampton). In 1992, two further UDCs was announced, one in Birmingham (Birmingham Heartlands DC), the other in Plymouth.

The UDCs were primarily located in inner city localities, yet, as Table 1.2

Table 1.1 Key dates in the lives of the Urban Development Corporations

Start-up date	Urban Development Corporation	Wind-up date
First Generation – 1981	London Docklands	March 31st 1998
	Merseyside	March 31st 1998
Second Generation – 1987	Black Country	March 31st 1998
	Cardiff Bay	March 31st 2000
	Teeside	March 31st 1998
	Trafford Park	March 31st 1998
	Tyne and Wear	March 31st 1998
Third Generation – 1988/89	Bristol	December 31st 1995
	Sheffield	March 31st 1997
	Central Manchester	March 31st 1996
	Leeds	March 31st 1995
Fourth Generation – 1992/93	Birmingham Heartlands	March 31st 1998
	Plymouth	March 31st 1998

Source: DETR, 1997

Figure 1.1 *Location of the Urban Development Corporations: England and Wales*

shows, there were great variations between the UDC areas (Urban Development Areas, UDAs) in terms of size and character. The UDAs ranged from Teesside with a massive 4,858 hectares to Central Manchester with just 187 hectares. Moreover, contrary to government assertions that UDCs were declared in areas that were essentially derelict, Table 1.2 shows that the majority of UDAs contained a significant industrial and employment base. As CLES (1990b) has noted, one can make a further distinction between UDAs which contained parts of the city centres, like Bristol, Merseyside, and Manchester, and those which were primarily comprised of industrial land, like Sheffield's Lower Don Valley, Trafford Park and Teesside. Plymouth UDC consisted of three sites in a single ownership, as opposed to Bristol where, as Oatley (1993) notes, part of the case for designating the UDC was the alleged fragmentation of land ownership.

There are, nevertheless, a number of features which were common to the UDCs (see, for example, Batley, 1989; Thomas and Imrie, 1997). In particular, central government, through the Secretary of State for the Environment (or the Welsh Office in the Welsh context), controlled key elements of the UDCs' organisation. As Figure 1.2 indicates, the Secretary of State, utilising the advice of consultants and other sources, defined the UDA and appointed a Board with responsibilities for appointing staff, devising strategy, and administering finance and resources. All UDCs, with the exception of Cardiff Bay, were responsible for development control powers, and could operate outside their immediate boundaries if they considered such actions to be beneficial to their regeneration plans. Their funding was gained through a mix-

Table 1.2 The UDCs at designation

Corporation	Size of urban development area (ha)	Population in the development area (numbers)	Employment in the urban development area (numbers)
Birmingham Heartlands	1,000	12,500	Not known
Black Country	2,598	35,405	53,000
Bristol	420	1,000	19,500
Cardiff Bay	1,093	5,000	15,000
Central Manchester	187	500	15,300
Leeds	540	800	NA
London	2,150	40,400	27,213
Merseyside	350	450	1,500
Plymouth	70	n/a	NA
Sheffield	900	300	18,000
Teesside	4,858	400	NA
Trafford Park	1,267	40	24,468
Tyne and Wear	2,375	4,500	40,115

Source: CLES, 1990b, and personal communications with the UDCs.

ture of three sources: an annual budget from central government; finance borrowed from the national loan fund; and the utilisation of receipts from land sales.

The UDCs also had the power to purchase land by agreement, to 'vest' it from public sector bodies, and/or to compulsory purchase it from private sector landowners. As Coulson (1990) notes, UDCs were dedicated bodies with the specific remit to secure land and property development, and, in this sense, they typified the property-led approach to urban regeneration. 'Development' was the key objective of the UDCs, yet its definitional basis was restricted to 'bringing derelict and/or vacant land back into productive

---

Secretary of State for the Environment, Transport, and the Regions

- Appoints Board
- Approves appointment of Chief Executive
- Approves staff terms and conditions
- Confers planning function

- Determines annual grant
- Authorises compulsory purchase
- Determines date of wind up
- Inherits residual assets and liabilities

Department of the Environment, Transport, and the Regions (and Government Offices for the Regions)

- Provide general guidance
- Designate Chief Executive as Accounting Officer
- Approve project expenditure above specified thresholds
- Monitor performance against corporate plan

- Set wind up objectives
- Provide wind up guidance
- Support legislative process

- Secure funding for residual liabilities

Urban Development Corporations

Board

- Appoint Chief Executive

- Set Corporate policy
- Determine planning applications
- Approve projects and expenditure
- Approve wind up plan
- Ensure wind up plan achieved

Chief Executive

- Responsible for financial management
- Ensures resources used efficiently
- Appoints staff
- Manages operations
- Manages wind up

Note: The Secretary of State for the Environment and the Department establish the administrative and financial framework for the Corporations. The Board are responsible to the Secretary of State for ensuring that agreed objectives are met. The Chief Executive is responsible for managing operations and is the Corporation's Accounting Officer.

Figure 1.2   *Key bodies involved in Urban Development Corporations*

Source: National Audit Office, 1997

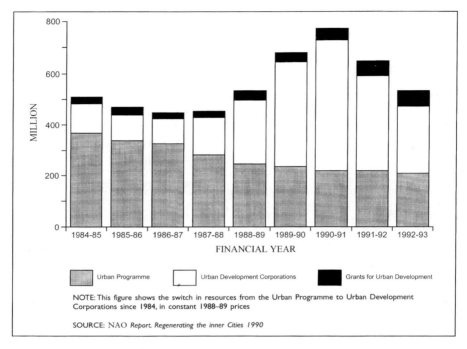

Figure 1.3 *Finances and resources*

use' (Local Government, Planning and Land Use Act, 1980, section 136). As Healey (1991) notes, the objectives of the UDCs were to achieve local economic growth by providing the physical infrastructures and locales appropriate for the new industrial and commercial sectors. In particular, the UDCs had the remit of unblocking supply-side constraints on the development potential of land and property, to assemble land parcels, develop infrastructure, and pump-prime land values in the inner cities as a means of attracting private sector investment.

It was always intended that UDCs operate as catalysts by providing a framework for private sector development interests, not as developers themselves (see Lawless, 1989). The UDCs were not the plan-making authority, and their development proposals had to, in theory, take account of local authority planning frameworks. Yet, as Lawless (1989), CLES (1992) and the case studies in this book indicate, the UDCs were never effectively bound by local planning frameworks, although the majority tried to operate within agreed strategies with the local authorities. A number of UDCs, including Trafford Park, Tyne and Wear, and Sheffield, drew up agency agreements enabling the respective local authorities to process planning applications on their behalf. However, as Byrne (1992) concludes, UDCs were able to effectively ignore the statutory planning process if they wanted to, and this certainly occurred in a number of the UDAs.

Table 1.3 Public expenditure (£ million) on the English UDCs (and other selected programmes) as a proportion of total expenditure on urban policy, 1991–92 to 1997–98

Expenditure Programmes (Selected)	1991–92	1992–93	1993–94	1994–95	1995–96	1996–97	1997–98
UDCs	508.6	430.6	343.1	258.0	217.9	193.8	168.0
English Partnerships	99.1	153.3	164.9	191.7	211.1	229.6	209.6
Housing Action Trusts	10.1	26.5	78.1	92.0	92.5	87.7	88.7
Challenge Fund	–	–	–	–	136.4	264.9	481.6
Estate Action	267.5	348.0	357.4	372.6	315.9	256.7	169.9
City Challenge	–	72.6	240.0	233.6	226.8	230.1	143.0
*Total Regeneration	1,096.8	1,456.3	1,616.4	1,517.0	1,347.7	1,434.9	1,369.9
Spend on the UDCs as a proportion of total spend (%)	46.3	29.5	21.2	17.0	16.1	13.5	12.2

*These figures also include expenditures on programmes not itemised above
Source: DETR, 1997

While the democratic credentials of the UDCs have been a recurrent concern to local politicians and communities, of equal controversy and debate has been the funding of the development corporations (see DoE, 1990). As Figure 1.3 shows, throughout the 1980s there was a shift in the balance of expenditure on the UP authorities and the UDCs (and other urban policies), with the Development Corporations becoming the priority funding initiative by receiving, in total, considerably more than the total grant for all the 57 UP Districts in England (CLES, 1990b). This shift was both symbolic and material in that it represented the decline in central government support for the formative (post-1968), local authority-driven, elements of urban policy, in favour of the UDCs property-based approach. Thus, by the early 1990s, as Table 1.3 shows, the UDC initiative was accounting for just under half of total government expenditure on the cities, a proportion that began to decline rapidly as new initiatives, such as City Challenge, came on stream (DoE, 1996a).

Over the lifetime of the UDCs, funding was highly variable. For instance, Table 1.4 indicates the privileged status of the LDDC compared with the other UDCs, in terms of receipt of grant-in-aid. Up to 1996, LDDC, for

Table 1.4 Grant-in aid (£ million) received by the Urban Development Corporations over their lifetime

Urban Development Corporation	Grant-in-Aid (£ million)
London Docklands	1860.3
Merseyside	385.3
Black Country	357.7
Cardiff Bay	519.7 (1996 estimate)
Teesside	350.5
Trafford Park	223.7
Tyne and Wear	339.3
Bristol	78.9
Sheffield	101.0
Central Manchester	82.2
Leeds	55.7
Birmingham Heartlands	39.7
Plymouth	44.5

Source: DETR, 1997
         CBDC, 1996b

example, had received public funding to the order of £1,860.3 million compared with the combined figure of £2058.7 million for all the other UDCs. In addition, the UDCs were expected to be profit-motivated, not only to break even on land deals but to engage in land transactions for profit, ploughing land receipts into development projects. Thus, an important source of revenue for the UDCs came from land sales, and prior to the property crash in 1989, the UDCs, especially the LDDC, were making significant profits on land purchase and sales deals. For instance, land sales by the LDDC raised £10 million in 1985–86, rising to £115 million in 1988–89. However, after the 1989 property crash, sales fell dramatically and many UDCs incurred debts on land purchase deals. For example, CBDC only generated £84,000 from land sales in 1989–90 while spending £1 million on land reclamation and £16.8 million on land acquisition.

Indeed, as Table 1.5 shows, many of the UDCs incurred liabilities at a time of falling land values, while committed to capital projects with few capital receipts to offset their costs. The implications for many UDCs were serious and for TWDC, for example, who spent £16 million on land purchase in 1988–89, the value of its land sales over the same period was only £2 million. It was not alone in facing this predicament. An analysis of government answers to questions posed in the House of Commons indicated that six of the UDCs had lost out on land purchases and sales. For instance, the Black Country showed a loss on property deals, based on March 1992 values, of £26.5 million over the period 1987–92. Likewise, Bristol, Central Manchester, Leeds, Trafford Park, and Tyne and Wear, incurred cumulative losses on land transactions of, respectively, £12.9 million, £8.3 million, £3.4 million, £10 million, and £6.3 million (in the period from their inception to early 1992).

The limitations of the UDCs general approach, in being dependent on rising (land) market valuations, intensified the fiscal pressures on their expenditure programmes. The collapse in property markets meant that by 1991 UDCs became even more dependent on government grants and public sector schemes to ensure that the momentum of regeneration was maintained (CLES, 1992). By the early 1990s, the UDCs were unable to dispose of land they had bought during the boom period of the mid- to late 1980s, and the underlying ethos of the UDCs, rapid development, was under severe threat. Indeed, as CLES (1992) has documented, a number of flagship projects were delayed and suspended, budgets and corporate plans were revised, while the collapse of Olympia and York in London Docklands still stands as a symbol of the vicissitudes of property-based regeneration.

While the revenue base of the UDCs was connected to prevailing market conditions, their expenditures reflected the property-led, leverage, role with the overwhelming proportion of their spending directed to promoting, even to the extent of directly subsidising, the development of land, but particularly through land acquisition, reclamation, and road building. For instance, the DoE (1990) estimated that of the total expenditure of the UDCs (excluding the LDDC) in 1990–91, 56% went on land purchase and assembly. Moreover, as Table 1.6 indicates, the dominance of land purchase and assembly, as the major item of UDC expenditure, was maintained throughout the life course of the second and third generation UDCs. Thus, 53% of their expenditure, or £399 million, comprised land purchases and reclamation, while a further £104 million (or 14% of total spend) was expended on transport and related infrastructure. In contrast, only 1% of total expenditure (or £8 million) was directed to housing and support to the community while £115 million, or 15% of the total budget, went towards administrative costs.

The spending programmes of the UDCs were, therefore, targeted at specific projects. For instance, evidence from CBDC, for the period April 1987 to March 1996, shows the weighting of its expenditure programmes primarily

Table 1.5 Selected UDCs' losses on land deals

Corporation	Amount spent on land transaction (£ millions)	Amount received from land transaction (£ millions)	Value of land 31st March 1992 (£ millions)	Loss (£ millions)
BCDC	68.0	3.7	38.8	26.5
BDC	24.4	nil	11.5	12.9
CMDC	12.3	nil	4.0	8.3
LDC	17.5	1.8	12.3	3.4
TPDC	41.6	9.3	22.3	10.0
TWDC	42.4	5.8	30.3	6.3

Source: Independent, 13th July 1992.

Table 1.6 Expenditure by the Second and Third Generation Urban Development Corporations, up to March 31st 1992

Expenditure Heading	Expenditure £ million (% of total spend)
Land purchase and reclamation	399 (52.9)
Administration	115 (15.2)
Transport and infrastructure	104 (13.8)
Support to industry	70 (9.3)
Environmental projects	57 (7.8)
Housing and support to community	8 (1.0)
TOTAL	753 (100)

Source: UDC annual reports and corporate plans

towards land purchase and assembly. As Table 1.7 indicates, 32% of its gross expenditure, that is £200,751,000, was incurred on a single project, the construction of a barrage across Cardiff Bay (see Chapter 4 for more details, also Rowley, 1994). In contrast, CBDC's spending on community programmes has been small, totalling nearly £19 million or 2% of its total expenditure in the period to March 1996. Such figures are more revealing when set beside other forms of expenditure. Thus, CBDC spent nearly 4% of its budget, or just over £23 million, on marketing alone in the period up to March 1996, while less than 1% was devoted to environmental improvement schemes.

Table 1.7 Expenditure on Selected Projects and Programmes by Cardiff Bay Development Corporation, April 1987 to March 1996

Expenditure Programme	Expenditure £'000 (% of total spending)
Barrage	200,751 (32.1)
Peripheral Distributor Road	29,011 (4.6)
Bute Avenue	20,709 (3.3)
National Techniquest	12,226 (1.9)
Cardiff Bay Retail Park	19,920 (3.1)
Inner Harbour	38,701 (6.2)
Marketing	23,298 (3.7)
Administation	47,914 (7.7)
Community Initiatives Programme	7,969 (1.3)
Environmental Schemes	4,753 (0.8)
Hamadryad Park	3,459 (0.5)
Mount Stuart School	2,665 (0.4)
Gross Expenditure	**624,129**
Grant-in-Aid	**519,735**

Source: Cardiff Bay Development Corporation, 1996b, Corporate Plan 8, November 25th

Grant-giving to the development industry was also one of the critical functions of the UDCs (see NAO, 1993).[5] As Table 1.8 shows, for example, the giving of grants was variable between the second and third generation UDCs. For instance, up to March 31st 1992, Central Manchester DC had allocated 39% of its grant-in-aid in the form of grants to private sector developers while, in contrast, Teesside had allocated 3%. Such variations are the results of an interplay of the differences between the prospects for profitable development in different UDAs and differences in the detailed approaches of UDCs to generating property-led regeneration. Yet, whatever the spatial variation, in terms of winners and losers in urban regeneration, the emergence of UDCs as a 'grant regime' seemed significant, particularly in directing resources to the development industry. Indeed, as Lawless (1991, p. 26) notes, 'the sector which benefited most from urban policy in the 1980's was the private sector in general and the development industry in particular'.

All of this sits very uneasily alongside other items of UDC expenditure. For instance, as Table 1.9 shows, UDC spending on community projects was small when set against expenditures on physical infrastructure. CLES (1992) comments that projects which had a social or community content were seen as a 'poor investment' by UDCs, primarily because their remit was to invest in schemes bringing in the largest amount of private investment at the lowest direct public subsidy. In contrast, UDC expenditure on marketing was seen as pivotal in order to sell development sites. As Healey (1992, p. 100) comments, the imagery and rhetoric of regeneration was seen as important, 'not just as political publicity, but to regenerate confidence in the property development possibilities of older industrial cities'. To such ends, the UDCs collectively spent £14.6 million, or 2.5% of total grant-in-aid received, on advertising and marketing in 1990-91, with the LDDC alone spending £2 million, or 0.6% of its allocated budget for that financial year. Between 1981

Table 1.8 Grant assistance as a proportion of UDC expenditure

Corporation	Expenditure to March 31st 1992 Grant (£m)	Expenditure to March 31st 1992 Total (£m)	Percentage of grant to total expenditure
Central Manchester	17	44	39
Black Country	19	143	13
Tyne and Wear	17	161	11
Sheffield	4	52	8
Trafford Park	8	100	8
Leeds	3	41	7
Teeside	9	177	5
Bristol	1	35	3
TOTAL	78	753	10

Note: Expenditure details are based on actual outturn each year.
Source: NAO, 1993

Table 1.9 Support to the local community as a proportion of UDC expenditure (% of total annual spend)

Corporation	1988–89	1989–90	1990–91	1991–92
BCDC	1.3	1.8	2.7	1.5
BDC	0.0	1.0	0.0	0.1
CBDC	0.6	0.9	1.0	NA
CMDC	1.4	1.6	3.4	2.3
LDC	0.9	0.1	0.2	0.2
LDDC	3.3	6.5	5.3	NA
MDC	4.5	3.1	4.2	NA
SDC	0.4	0.7	1.0	1.4
TDC	2.4	0.3	0.2	0.9
TPDC	0.1	0.3	0.4	0.3
TWDC	0.3	0.9	1.3	1.0
Average	1.4	1.5	3.7	1.0

Source: Hansard, 1989, Annual Reports and Accounts of the UDCs (1990–91), NAO, 1993.

and 1991, the LDDC spent £20 million on promotion and publicity (CLES, 1992).

While controversies will remain concerning the appropriateness or not of UDCs' spending patterns, sensitising themselves to their locales was also a priority of the UDCs given the recognition (by many development corporations) that progress would be easier with the co-operation of local actors, agencies and the wider range of community interests. This conclusion was reached by LDDC after eight years of protracted conflict with the local authorities, though, as Brownill points out in Chapter 2, the increased sensitivity to community and local authority interests was born of expediency, not altruism, and was reversed when political conditions changed. In addition, the London experience, of conflict leading to almost total exclusion of local authorities from the regeneration process, seemed to have cautioned most local authorities against too critical a stance in relation to UDCs in their areas. As a result, the House of Commons Employment Committee (1989, p. 6) reported that the newer UDCs had 'the full co-operation of the local authorities'.

Since the first phase designations, subsequent UDCs developed a softer, more conciliatory, approach to local consultation by devising closer links with a range of community and local organisations. The second wave of UDCs coincided with a reappraisal of the nature of UDC strategy and policy with a number of influential bodies criticising their focus purely on physical regeneration. For instance, the House of Commons Public Accounts Committee (1988, p. 8), while acknowledging the physical achievements of the LDDC, noted that 'UDCs should seek from the start to strike a reasonable balance between the physical development of their areas, and the social and other needs of those living there'. Echoing this, the House of Commons

Employment Committee (1989) concluded that the UDCs should adopt a more precise definition of regeneration, including employment and unemployment objectives. The Committee noted that UDCs 'should be charged with greater responsibility for ensuring that communities directly affected by them, and in their neighbouring areas, benefited from regeneration' (1989, p. 2). It was concluded that, 'UDCs cannot be regarded as a success if buildings and land are regenerated but the local community are bypassed and do not benefit from regeneration' (1989, p. 3).

These comments reflected an amalgam of community and political pressures on UDC programmes to widen their scope, yet only in very few of the UDCs did community and social issues feature as an integral component of their strategies from the outset. For instance, the BCDC, TWDC, SDC and CMDC all appointed community liaison officers, while TWDC appeared to be one of the more progressive in preparing a Community Development Strategy while recognising the need to obtain community gains from property developers. As Robinson and Shaw (1991) note, the more sensitive second and third generation UDCs undoubtedly paid lip service to the community, while falling far short of a genuine democratisation of urban policy. Yet, it must also be acknowledged that local party political processes are also exclusionary, and Brownill *et al.* (1997) have argued that the arrival of UDCs, with their need for some kind of democratic legitimation, provided ad hoc opportunities for influence by groups of residents who had little chance of a voice in mainstream political life.

Structural constraints on the UDCs' capacity to operate are rarely mentioned in evaluative literature, while some writings still have a tendency to characterise them as institutions 'outside' of their localities, autonomous islands of development, with few, if any, institutional linkages with local politicians and community groups. But, as CLES (1990b), and others, indicate, a wide range of institutional linkages characterised UDC-local authority relations, from joint forums, and budgets, to a wide range of informal meetings at member and officer level. These linkages suggest, that while UDCs had a range of similar powers, their actual practices, strategies and approaches to regeneration varied considerably. Some of the more obvious continuities and differences in the successive generations of the UDCs have been discussed by Stoker (1989). In particular, Stoker notes how the second generation was less conflict-ridden than the first, partly because of the transformed political climate of the mid- to late 1980s where pragmatism, or what Stoker refers to as a 'watchful co-operation', became a defining feature of UDC-local authority relationships.

Moreover, contrasts between the different styles and policies of the UDCs is illuminating in highlighting the diverse approaches underpinning strategies of property-based regeneration. While many accounts portray UDCs as invariant institutions, Stoker (1989, 1991), and others, note how the specificity of local politics, and the diverse material conditions in the UDAs, led to significant contrasts in the strategies and policies of the UDCs. For instance, in the field of equal opportunities, SDC was unique in monitoring the ethnic background of applicants who submitted planning applications, largely because of the agency agreement it had with the city council in relation to development control. Likewise, CBDC monitored who was getting what jobs and also operated training initiatives and work schemes for locally unemployed people. Moreover, BDC signed a 'Concordat' with the city's Race Equality Council on the necessity to secure gains for black and ethnic minorities in the city. Oatley (1993), however, has cast doubts on the value of the Bristol Concordat but concedes that the BDC had to make some moves to forge links with local groups and agencies.

The range of institutional networks, between the UDCs and actors and agencies, was much greater than is supposed, while it is increasingly clear that the property-led objectives of the development corporations did not remain unaffected by alternative, sometimes competing, policy goals which emanated from local sources (Clavel and Kleniewski, 1990). This is certainly one of the observations by CLES (1990b), yet there is still an issue of how far, and in what ways, a pluralist policy system was able to emerge and operate under conditions largely set by the central state, an issue which the contributors to this book address in some detail.

Assessing and evaluating the Urban Development Corporations

There is a paucity of published evaluations of the performances of the UDCs. What exists is a range of disparate writings, some government commissioned, others independent academic research, with the focus often on single cases or specific aspects of UDC operations (Imrie and Thomas, 1992; NAO, 1993, 1997; Thomas and Imrie, 1997; Williams, 1994). The overall effect, then, is partial and limited coverage with fragmented insights into the operations, practices and policy effects of the UDCs. In particular, the DETR's internal evaluations of the UDCs have always revolved around a narrow range of criteria as indicated in Table 1.10. As the table shows, DETR's assessment of the UDCs is primarily focused on quantitative measures of the 'bricks and mortar' (literally) laid down in the UDAs (also, see, DoE 1996b and c). For some commentators, such evaluative criteria are narrowly conceived and are characterised by inconsistent and vague definitions (CLES, 1990b; Imrie, 1996; Turok, 1991). In particular, CLES (1990b) criticise UDC output measures for failing to provide clear, consistent, and standardised information. As they have argued:

> The confusion and incoherance of figures provided in the DoE and UDC annual reports is even more galling when one recognises how little information is made public about UDC performance. On the rare occasions that public accountability can operate, the result is the provision of information that is so confused and unreliable that it often beggars belief. (*Ibid.*, p. 14)

Table 1.10 Attributed achievements of the UDCs up to March 1997

Urban Development Corporation	Land Reclaimed (hectares)	Housing Units Completed	Non-Housing Completed (000 sq. m)	Roads, built or Improved (km)	Gross Gain Permanent jobs	Private Sector Investment Committed (£m)
London	776	2165	2300	282	70484	6505
Merseyside	382	3135	589	97	19105	548
Black Country	363	3441	982	33	18480	987
Cardiff Bay	310	2260	378.5	26.6	9387	774
Teeside	492	1306	432	28	12226	1004
Trafford Park	176	283	636	42	23199	1513
Tyne and Wear	507	4009	982	39	28111	1115
Bristol	69	676	121	6.6	4825	235
Sheffield	247	0	495	15	18812	686
Central Manchester	35	2583	138	2.2	4944	373
Leeds	68	571	374	11.6	9066	357
Birmingham Heartlands	115	669	217	40.8	3526	217
Plymouth	10.8	28	3.5	4.4	29	3.9
TOTALS	3553	40576	7650	628.2	222194	14319

Source: DETR, 1997; Welsh Office 1997

Figures for Cardiff Bay are up to June 1997.

London docklands includes all infrastructure i.e. Docklands Light Railway, footways, cycleways and utilities.
Merseyside figures include water areas.
Black Country includes £314 million investment on fixed plant and machinery.

DETR's evaluations of the UDCs have tended to reflect the pragmatic concerns with costs and differentials between inputs and outputs. Their primary concern is with efficiency evaluation or the capacity of organisations, such as UDCs, or policy activities, such as land disposals, to produce measurable results. Not surprisingly, then, evaluations are performance-related and the criteria used are limited in range, such as jobs created and safe-guarded, hectares reclaimed, and quantities of constructed roadway. Such criteria are usually reinforced by a concern with the economic efficiency of programmes, including maximising output while reducing resource expenditure. The emphasis is 'value-for-money' which is a catch-all phrase denoting that some demonstrable (quantifiable) return must be made on public investment. However, some commentators, while recognising the value of DETR's standard measures, have argued that a greater range of (non quantifiable) variables ought to be accounted for in measuring and evaluating the performances of urban policy programmes (Hambleton and Thomas, 1995; Imrie, 1996; Turok, 1991).[6]

McAllister (1980), for example, notes how the subjectivities of people and places should be integrally interrelated to evaluative indicators, while evaluators should not assume that their values are a good indicator of the values held by the people that they purport to serve. For McAllister (1980), then, there is a need to develop evaluative criteria by asking the recipients of urban policy, that is, the local communities, what they regard as of value to them. Likewise, Imrie (1996) has called for a transformation in the social relations of policy evaluation to permit local communities greater involvement in the design and evaluation of urban policy. Such observations are driven by a recognition of the inability of standard evaluative techniques to differentiate between the diverse needs of community groups, a situation illustrated by the reliance on aggregate indicators of performance. Yet, as Turok (1991) notes, this often ignores special needs of particular groups while it is clear that some evaluative values, especially those related to deep moral feelings, cannot be easily conveyed by DETR's reliance on a quantitative approach.

The kinds of problems identified by Robson *et al.* (1994, p. 437– 440), in relation to urban policy in general, also ring true for UDCs. Foremost, evaluation and serious independent monitoring has not been an organisational priority of UDCs or central government departments. As Brownill *et al.* (1997) have argued, UDCs are 'can-do' organisations which put their efforts and money into 'getting things done', not evaluating impacts. Consequently, the availability of baseline and longitudinal data, to name two components of any kind of evaluation, has been variable – both spatially and across topics. Other than DETR, the most comprehensive government evaluation to date of the development corporations is the NAO (1993) report on the achievements of the second and third generation UDCs. In places, the NAO report is hard hitting in criticising the internal monitoring standards of the UDCs and, as it stated, 'UDCs do not keep explicit summary information on the progress of individual projects to time and cost and this makes it difficult to arrive at a firm view on their project control' (p. 3). The NAO also concluded that about 60% of the UDCs annual performance targets in key areas had been met although, as the report noted:

> the National Audit Office had reservations on such matters as the definition of performance measures, the achievability of targets and the reliability of output data. It was, therefore, difficult to establish the strength of each corporation's performance and the value for money achieved. (*Ibid.*, p. 1–2).

It is clear that neither the will nor the data exist to undertake convincing evaluations of the impact of the UDCs which take account of additionality, deadweight, and other complexities which must be addressed in serious evaluations of urban policy (see Hambleton and Thomas, 1995). Perhaps most serious of all has been what Robson *et al.* (1994, p. 438) call the general lack of independent local watchdogs capable of collecting and analysing their own data in order to construct regular and comprehensive evaluations of the UDCs. Even critical commentaries and assessments of the UDCs have tended to rely on data collected by the Corporations themselves or by agencies commissioned by them to do so.

Taken at face value, however, DETR's data demonstrate some significant achievements by the UDCs in physically transforming parts of the British cities (Table 1.10). For instance, the figures attribute LDDC with the reclamation of 776 hectares of derelict land, while encouraging housebuilders and housing associations to complete 21,615 housing units in London docklands. Likewise, the data suggest that MDC reclaimed 382 hectares of derelict land in Liverpool while securing 19,105 (gross) jobs over its lifetime. Similarly, Teesside, perhaps the most property-focused UDC, reclaimed 492 hectares of derelict land, while claiming to have created 12,226 (gross) jobs. In total, the UDCs, up to March 1997, had reclaimed 3,553 hectares of derelict land, while presiding over the creation of 222,194 (gross) jobs. In addition, their programmes had facilitated the building of 40,576 housing units and 7,650 square metres of non-housing units (i.e. primarily office space). However, the UDCs failed, in DETR's terms, to secure the ratio of 1:4 public to private investment which was part of their remit. Thus, as Table 1.11 indicates, in aggregate, the UDCs achieved a ratio of public to private investment of 1:3.6, although some of the UDCs, such as Sheffield and Leeds, achieved well above DETR's targets.

Whatever the physical targets attained by the development corporations, they have been criticised by all sides of the political spectrum, by the left for

Table 1.11 Ratio of public to private investment in the UDAs, up to March 1997

Urban Development Corporation	Grant-in-Aid (£'000)	Private Sector Investment Committed (£'000)	Ratio of public to private investment
London Docklands	1860.3	6505	1:3.4
Merseyside	385.3	548	1:1.4
Black Country	357.7	987	1:2.7
Cardiff Bay	370.0	774	1:2.1
Teesside	350.5	1004	1:2..8
Trafford Park	223.7	1513	1:6.7
Tyne and Wear	339.3	1115	1:3.2
Bristol	78.9	235	1:2.9
Sheffield	101.0	686	1:6.7
Central Manchester	82.2	373	1:4.5
Leeds	55.7	357	1:6.4
Birmingham Heartlands	39.7	217	1:5.4
Plymouth	44.5	3.9	1:0.0
Totals	3918.8	14319.6	1:3.6

Source: DETR 1997
      CBDC 1997b, Welsh Office, 1997

Figures for Cardiff Bay up to June 1997.

circumventing local democracy and asserting the primacy of market goals over social and community objectives, and by the right for running projects and programmes seen as wasteful expenditure, while replicating much of the bureaucracy they were supposedly put in place to overturn. Indeed, one of the main concerns has been job creation, yet reviews of the UDCs suggest that their strategies have been less than successful in translating investment into jobs, while the much vaunted trickle-down has been conspicous by its absence (National Audit Office, 1988).[7] As the NAO (1988, p. 4) commented on the Merseyside DC, 'their job creation programmes have only had marginal impact', yet, as one Chief Executive of a UDC has argued, 'we were not given a brief to create employment'. Indeed, the DoE (1988, p. 4) reinforced this view in saying that, 'we do not see UDCs as being primarily or immediately concerned with employment; they are about regeneration, and indeed the physical regeneration of their areas'.

Yet, even in physical terms, the success of the UDCs has been questioned, and many consider that what has been built in the UDAs would have been constructed anyway; that the designation of the UDCs has made only marginal difference to the investment climate (Turok, 1991). While this is a contestable statement, and difficult to evaluate, the property crash 1989–92 fuelled political debate concerning the status of property-led regeneration and, as CLES (1990a and b) indicated, the majority of UDCs were slow to show any results. The property-led approach of the UDCs also led to the charge that there was a lack of concern for the strategic implications of UDC activity (Imrie and Thomas, 1992). CLES (1990b) argued that, in spite of UDC rhetoric about strategy, the view of local authorities was that UDC policies and programmes were not strategic in looking at city or community-wide implications. A concern for a comprehensive strategy within its boundaries – as in the case of CBDC – did not imply a great concern for implications outside the boundaries (though, in CBDC's case, it was put under pressure to jointly fund studies on the implications of its policies for public transport links to the city centre and city centre retailing).

The absence of a strategic perspective, in seeking solely to regenerate without due regard to people and places beyond the boundaries of their UDAs, was also evident elsewhere. For instance, a judgement handed down by the High Court in May 1997 castigated TDC for ignoring a number of material considerations, including the Cleveland structure plan, in granting a planning permission to Asda to develop a superstore at Middlehaven, one of the locales within the UDA. In throwing out TDC's permission, the presiding judge argued that the corporation had, from an early date, 'elevated the regeneration potential of the proposal to a level at which objective judgement of its planning merits was foreclosed' (Bond, 1998, p. 13). TDC was particularly criticised for ignoring the spirit of Planning Policy Guidance which states that edge of centre sites should be within 200 to 300 metres walking distance of town centres. However, the proposed Asda development was miles out of town, while Cleveland Borough Council and Morrison's, the supermarket chain, also objected to the scheme on the basis of it detracting from existing retail provision within the locality.

Such evidence suggests that the seeming single-mindedness of the UDCs, much lauded by Heseltine in 1979, failed to deliver the expected returns. Indeed, it was this single-minded approach, the avoidance of local government bureaucracy, which was held up as the key to UDC success. Yet, one of the ironies of the whole programme was the bureaucratic and legal ties on UDC operations emanating from central government, and, as the following chapters will show, UDCs were far from the *laissez-faire*, debureaucratised organisations that they were purported to be under the 1980 Act. For instance, DoE guidelines prevented UDCs from making payments in excess of £1,000 without prior government consent, while one of the UDCs most significant spending headings was 'administrative and related costs'. Moreover, all land purchases by the UDCs had to gain prior approval by central government, and, as CBDC commented, 'while we want to buy up land at the market rates central government won't let us . . . they want it on the cheap' (Imrie, 1993).

While debates will continue about what the UDCs did and did not achieve, at the time of writing all Corporations bar CBDC have been wound up, handing back responsibilities for their development areas to a variety of agencies and organisations. This process of exiting has, of course, been known about from the outset, since UDCs have always been fixed-term agencies. Yet exit strategies have been fraught with difficulties and a number of commentators have noted how DETR has given insufficient attention to some of the political, fiscal, and economic implications of the process (Oatley, 1998; also refer to Chapter 8). The process of exit was never considered in the original 1980 framing legislation and when guidance was issued in 1992 it conceived of the process as linear, orderly, and rational with sufficient time and organisational capacity for agreement to be reached between the participating parties. Indeed, as Table 1.12 indicates, the DETR considered a two-year timetable to be appropriate for wind up with the NAO (1993, p. 34) concluding that 'the Department and the Corporations have taken steps to organise their affairs in an orderly way to prepare for wind up'.

The reality of exit is quite different to that portrayed in the DETR's idealised version of the process (also see Oatley and May, Chapter 8).[8] In the case of London Docklands, for example, LDDC has left behind a complicated tangle of assets and liabilities in the Royal Docks to be handed back to the local authority, Newham Borough Council. For instance, Newham is adopting the Royal's road network but, as Bond (1998, p. 13) suggests, while it is only five years old it is showing its age and will need substantial resources to maintain it. LDDC has also left 120 acres of land in the Royal Docks still to be developed and most of this will be taken over by another quango, English Partnerships (EP), who intend to spend £100 million there over the next ten years. Likewise, when EP hands back Thames Barrier Park to Newham it is estimated that its upkeep will cost Newham £700,000 over

Table 1.12 Possible timetable for wind up of a UDC

24 months prior to Exit	Wind up date announced
24 to 18 months prior to Exit	Firm up proposals internally and at UDC Board level. Discussions with DETR and Treasury
12 months prior to Exit	Successor arrangements confirmed. Planning reverts to local authorities
12 to 6 months prior to Exit	Negotiations with local authorities and other agencies. Staff interviewed
6 months prior to Exit	Individual staff informed of futures
6 months up to Exit	Detailed arrangements put into place. Staff counselling and assistance
Exit	Successor body takes over
Up to 6 months after Exit	Finalisation of accounts

Source: Adapted from NAO, 1993

six years (Bond, 1998). EP's involvement in London Docklands is indicative of the emerging system of governance within some of the UDAs which is not very different to what has gone, inasmuch as public sector organisations, with no direct local accountability and a narrow focus on property development, will continue to play dominant roles.

Thus, the Commission for the New Towns is picking up the assets and liabilities of the UDCs while EP, as already intimated, is assuming the regeneration role in some unfinished projects. For instance, EP will assume control over the redevelopment of the Royal William Yard scheme in Plymouth while, in Tyne and Wear, it will invest £13 million in a range of projects including the TWDC initiated Viking industrial park in South Shields. Some local authorities appear wary about the new arrangements and as Councillor Conor McAuley, from Tyne and Wear, has argued:

> to date English Partnership has been fairly open about wanting to work with us and we welcome that but we've not seen any evidence that they've actually got their act together yet. (Quoted in Bond, 1998, p. 14).

When EP does 'get its act together' neither it, nor its partners, will have the benefit of a comprehensive evaluation of the UDC experience on which to build, but, as we have demonstrated, there is no shortage of better and worse informed judgements of what UDCs have achieved, or failed to deliver. The controversy in which UDCs were conceived and operated looks set to continue as contested evaluations of their legacies. In the next section we consider one debate of especial significance to our understanding of urban policy – namely the extent to which individual corporations need to be understood as agencies embedded in local political and economic linkages.

Urban policy, locality, and the UDCs

Despite the considerable volume of academic research and reporting on the

UDCs, relatively few attempts have been made to document and assess the range of UDC policies and practices (although, see Thomas and Imrie, 1997). The UDCs have been submerged under stereotypical conceptions which variously portray them as 'executives' of the central state, 'puppets' for global corporate capital, and mechanisms for overriding local democratic institutions. Yet, as the chapters in this book will go some way to illustrating, no single UDC rigidly conforms to the contours outlined above, while it is clear that there are significant variations in their modes of operation (see Thomas and Imrie, 1997). In particular, it is increasingly clear that there is imperfect understanding of the localised development and delivery of national urban policy, or of how its wider development objectives are sustained, modified, or contradicted, by local socio-political milieux.

Thus, the starting point, common to all contributors, is the likelihood of significant variation in the delivery and implementation of urban policy, its unevenness, whatever the stated objectives of national government. In particular, this perspective, a need to acknowledge contingent relations, layers of local, or sub-national, political and institutional autonomy, seems self-evident given the very different socio-economic and political histories of the UDC localities, legacies which have been fundamental in shaping the precise configuration of (local) policy content and implementation. In making the commitment towards a more sophisticated view of the UDCs, the contributions in this book signal a discontent with policy studies literature which takes agencies and institutions out of the analysis, while reading off urban development as, in particular, a part of a uniform (economic) global logic. The totalising nature of such conceptions tends to ignore the general point that all social phenomena have causal powers at whatever spatial scale they are identifiable. Indeed, the UDC initiative, far from emasculating local political autonomy and responsibility, has been part of a wider process of institutional changes which have redefined the roles of local authorities inside the state.

Yet, there are a number of senses in which the themes of the book also go beyond the local in discussing some of the major structural influences on the content and scope of local actions. Echoing Logan and Swanstrom (1990), any contextual understanding of the UDCs, and urban policy as a whole, requires some discussion of how local actions are fashioned by wider structural opportunities and constraints. In particular, the book identifies a number of powerful structural forces which have influenced, and which continue to influence, the content of urban policy. Foremost is the embeddedness of cities in central state structures which clearly define the limits of local government autonomy, and demand recognition that the policies of the central state are a major player in the content of local policy formation and implementation. In recent political history, for example, the poll tax fiasco highlighted the authoritarian centralism of the British state, while legislation, like the 1989 Local Government and Housing Act, has only served to dilute once taken-for-granted local government powers.

Moreover, Sassen (1991) and others stress the importance of international, global, economic and political forces in determining the trajectory of urban development. Indeed, the globalisation of financial services, coupled with new technological capabilities, has only served to speed up flows of capital into, and out of, the world's cities. As Amin and Robins (1991, p. 28) note, if we are to consider 'that this global arena is shaped and informed by formidable relations of power, then the scope for local autonomy becomes considerably narrowed'. However, the book also tries to challenge the idea of a purely economic imperative underpinning UDC policy. While conservative ideologues have been calling for increased competition between cities, adapting to the imperatives of economic restructuring and the market, the contributions in this book echo Logan and Swanstrom (1990) in questioning the view that there is a market logic to capitalism to which urban policy (including the UDCs) at all levels must submit. In this, we concur with Walton (1990) who notes that both neo-liberal and neo-conservative theory have produced erroneous theoretical accounts of urban change, comprising a reification of markets or structures, a neglect of agency, and a failure to recognise or explain variation in the patterns of urban policy and performance, whilst, simultaneously, legitimising growth policies over social redistributive, or welfare, goals.

In developing some of these conceptual points, the original remit of the book was the concern with the delivery, development, and implementation, of UDC strategies and policies. In particular, we collectively identified three significant issues which have provided the focal points for the case study chapters in Part Two of the book:

(1). How did UDCs develop and formulate policies and programmes in relation to the opportunities and constraints of their locales? UDCs were never passive institutions that imposed solutions or remained unaffected by their localised operational environment. Much of the theoretical literature is increasingly sensitive to how external pressures are mediated by individual agents (Goodwin, 1991). Healey and Barrett (1990) have identified the need to give more attention to the way individual firms and agents interrelate in the negotiations of particular development projects and how, through these transactions, land and property markets are constituted and built environments made. In particular, the renewal of many UDAs rejects a commonly held view that UDCs were seeking, as a deliberate aim of policy, to exclude both local government and local interests from its legitimate functions. Such crude instrumentalism is questioned by the range of chapters in this volume which reveal the development of a more flexible approach towards policy formulation and delivery, one which encouraged pro-active inter-agency involvement. This reveals the limitations of theoretical positions which exclude the possibilities of local modifications of national urban policy (Healey and Barrett, 1990).

(2). What was the nature, extent, and influence of institutional innovation, interaction, and collaboration, between the UDCs and local participants in development? The actions of UDCs represented the implementation of a strand of modified national urban policy. The factors which influenced its impact undoubtedly included the local institutional milieux, local political systems, and the interactions between local and national actors. In this sense, evolving strategy was a hybrid, or amalgam of interests, characterised by new institutional forms and relations to those which traditionally deliver and formulate urban policy. This perspective contrasts with those who present the UDCs as institutions which remained largely unaffected by their local operating contexts, almost, at will, able to impose national policy guidelines and strategies. Yet, as the chapters in this book document, many of the UDCs were actively involved in complex inter-institutional relations at the local level, depicting a breakdown in some of the real barriers of (perceived) prejudice between the private, public, and voluntary sectors.

(3). What are the emerging legacies of the UDCs in terms of their influence upon local governance, policy styles, and the social, economic, and physical fabrics of the localities they operated within? How have the practices of the UDCs influenced contemporary urban policy and do the UDCs have anything of positive note to bequeath to future policy? As the evidence from the book suggests, the UDCs were part of an evolving system of local governance, both contributing to, and being influenced by, changing political and institutional responses in cities to economic restructuring and a neo-liberal central government policy agenda. The case studies in the book show quite clearly how mistaken it is to regard the longer-running UDCs (especially) as invariant over time. These were organisations which, themselves, changed, to varying degrees, in terms of policy content and style of operation. Consequently, any discussion of the local legacies of UDCs must not operate with a crude model of the 'UDC approach' and simply see whether it lives on or not. Some UDCs, it appears, did operate in a fairly crude way – see Robinson *et al.* on TDC, in this book – and their local legacies are much easier to calculate. But Deas *et al.* provide a picture of CMDC's having a complex role in Manchester; and their theme of the creation of UDCs as allowing new policy directions and/or modes of governance to crystallise is picked up, in a slightly different fashion, by others, such as Thomas and Imrie's discussion of CBDC.

Part Two of the book comprises eight studies of the policies, programmes, and operations of the British UDCs. In Chapter 2, Brownill emphasises the difficulty and danger of producing a simple evaluation of LDDC, an organisation which spent enormous sums of money over a 17-year period in a socially and economically complex area. Its impact varied spatially and sectorally, and its mode of operation changed over time. It significantly transformed the socio-physical and economic structure of Docklands yet, as

Brownill notes, it failed to address the problems of poverty and disadvantage amongst many of the communities within its jurisdiction. By way of contrast, Meegan's wide-ranging account of the MDC in Chapter 3 develops the theme that the early period of contested governance on Merseyside has been evolving into a more consensual one with the MDC making a significant contribution to this development within changing local and national political circumstances. For Meegan, MDC was never a full-blooded entrepreneurial organisation engaged in place marketing, for the political and economic circumstances of Merseyside did not allow it to be.

In Chapter 4, Thomas and Imrie show how CBDC gained local political and community acceptance by its incorporation into a pre-existing political consensus revolving around the idea of modernisation. By integrating itself into the locale, extending and reworking many of the widely held, and local, convictions concerning spatial development, CBDC did much to diffuse potential opposition while assuming a powerful role in determining the spatial trajectory of the city. In contrast, Byrne, in Chapter 5, develops the theme of how the material, cultural, and symbolic significances of places were being ignored by the seemingly crude approaches of the UDCs. In the case of TWDC, Byrne compares it with a form of colonial administration, an organisation wholly inappropriate to the task of maintaining the industrial and maritime heritage of the locality. As Byrne notes, TWDC's exit strategy was a continuation of its 'anti-industrialism' or a series of post-UDC development strategies which will continue, in his estimation, to undermine the fabric of labour and locale within the north east of England.

Robinson, Shaw and Lawrence, in Chapter 6, consider TDC to be the classic illustation of what was problematical about the wider Thatcherite agenda for the inner cities. As their chapter persuasively demonstrates, TDC brought much needed investment to a poor area but put most of it where it was not needed, and on projects which did little to address problems of poverty and marginality. For the authors, TDC can only be understood from a state-centred approach which sees 'quangos as executives of the central state, geared to regaining control at the local level by undermining local democratic institutions'. Dabinett and Ramsden, in Chapter 7, develop the argument that while the SDC achieved a significant regeneration of the Lower Don Valley in drawing in major investors, it failed to improve adequately the quality of life of people who experience disadvantage. For Dabinett and Ramsden, 'all projects funded out of public sector resources made available through urban policy should be able to illustrate clear and substantial benefits to the disadvantaged groups in the city region'. As they argue, SDC was limited because of a reliance on flagship projects when what was required was an urban policy for people 'smaller in ambition, broader in scope, and more neighbourhood based'.

In Chapter 8, Oatley and May outline the case of BDC, developing the argument that its approach was wholly inappropriate to the problems of Bristol's inner city. Throughout its life, BDC was seen by local politicians and community groups as an unacceptable imposition which, in the words of Oatley and May, paid little attention to the local democratically elected authority or to neighbouring communities. Oatley and May discern little contribution being made to the development of urban policy by BDC – the corporation remained fixed on its own agenda and was perceived as extremely instrumental in its dealings with local agencies. In contrast, Deas *et al.*, in Chapter 9, note that the imposition of CMDC was a pivotal moment in the emergence of a new entrepreneurial politics in Manchester. As they argue, CMDC played an important role in what they term the 're-corporatisation' of local governance in Manchester and the authors see the UDCs as much more than about physical and economic regeneration but as key agencies in the vanguard of transforming 'institutional relationships and established outlooks on economic development'.

In Part Three of the book, the chapters reflect on the lessons to be learnt from the UDCs while discussing the emergent urban policy frameworks seeking to supplant the development corporations. In Chapter 10, Colenutt notes that regeneration for people was not the main priority for successive Conservative governments during the 1980s and 1990s. Towards the new millennium, Colenutt is encouraged by a new climate of socio-cultural tolerance and the resurgence of regionalism and localism. In particular, in a context of fragmented, and fragmenting, local governance, Colenutt is more optimistic about the possibilities for community involvement in influencing the contours of urban policy. In the concluding chapter, Cochrane provides an assessment of the legacies of the UDCs. He acknowledges that they represented a break from previous forms of urban policy because of their single-agency, dedicated, approach. He also notes that the UDCs, perhaps paradoxically, were in the vanguard in encouraging the development of public-private partnerships. Cochrane also conceives of the UDCs as 'symbols of managerialism rather than democratic accountability' and, in this sense, they were, he argues, part of a much broader shift in governance structures in the UK and beyond.

In terms of the broader lessons to be learnt from the UDCs, the chapters highlight a number of important themes. Foremost, various contributors contend that the activities of the UDCs demonstrate the importance of the public sector in underpinning the regeneration of cities. Indeed, as Meegan in Chapter 3 notes, 'the overriding message of the MDC's activities on Merseyside has to be the continuing need for a massive, and sustained, public sector-led intervention in the regeneration of such disadvantaged city-regions'. Likewise, Dabinett and Ramsden make a point of contrasting the certainty of public sector contributors with the insecurity associated with private finance. In Sheffield, 'most major initiatives ended up being paid for by the public sector'.

Some of the chapters also question the extent to which UDCs meshed with local interests and gradually became embedded within their localities. For Robinson, Shaw and Lawrence, in Chapter 6, the Teesside experience 'suggests that this UDC consistently served as an arm of a national (unreconstituted Thatcherite) urban policy . . . and acted to exclude local government and other local interests from involvement in the process of regeneration'. Byrne's analysis of TWDC portrays the UDC's isolation and estrangement from the locale. However, other contributors note how local pressure, from councils, community groups and business associations, also encouraged, or even forced, some development corporations into sensitising their policies to local circumstances. Some of these, such as BDC's concordat with the Bristol Race Equality Council, have been dismissed as tokenistic, but there is evidence of some significant initiatives in addressing local employment and community concerns in some of the UDAs (see, for example, the Cardiff case in Chapter 4).

The variety in the nature and extent of relationships between UDCs and other agencies is a striking feature of the case studies which follow. In neither Leeds, Sheffield nor Bristol was a development corporation sought, yet, in Leeds, the UDC was absorbed into an inter-agency network which became a model of co-operation (Roberts and Witney, 1993). As Roberts and Witney (1993) note, the absorption of Leeds DC into the local political milieu was facilitated by pre-existing corporatist politics in the city. In Liverpool, by way of contrast, the political control of the City Council, by Militant sympathisers in the early part of the 1980s, was such that MDC could do little to find common ground even if it had wished to do so. However, as Meegan makes clear, the Corporation broadly followed established structure plans and policies and by the end of the 1980s, in a context where the local politics of Liverpool had moved more to the centre, MDC was beginning to develop partnerships and open up co-operative ventures with the local authority.

One of the alleged policy effects of UDC policies was their encouragement of social and spatial polarisation through gentrification. By the mid-1980s, there were outpourings of literature castigating the UDC-led yuppification of the urban development areas (Short, 1989). However, such portrayals were based on partial or select experiences and never captured the complexity of social changes occurring in the different UDAs. Indeed, some of the contributors to this volume note that the Development Corporations have been important in contributing to aspects of sustainable urban regeneration in encouraging the use of, for instance, brownfield sites for housing. For example, in Chapter 3, Meegan highlights the MDC's successes in populating derelict parts of inner Liverpool. He also points towards the creation of urban villages in the city which are helping to retain working class communities. Dabinett and Ramsden also acknowledge that the Lower Don Valley in Sheffield has been transformed by the activities of the SDC. Thus, they document the success of Meadowhall, a major shopping centre, and the associated leisure facilities. Significant new investment has also been attracted including back offices of the Halifax bank and an Abbey National shareholding centre. Similar evidence of physical transformation can be cited for London and Cardiff.

The lack of a strategic approach to urban regeneration is a recurrent theme about the operations of the UDCs. The desire to attract private sector investment and be seen to be successful over fairly short periods of time created a presumption in favour of accepting/promoting development of almost any kind, in the short term, and worrying about strategic consequences (if at all) later. The London Docklands experience – especially in the early years – was an extreme example of this, but there were clear echoes elsewhere, as virtually all the case studies in the book make clear. It might, at first, have seemed reasonable to hypothesise that smaller UDCs would also be less concerned about the impacts of developments they promoted on areas outside their boundaries; however, the evidence of the case studies is that even larger UDCs (such as TDC, in Chapter 6) could be as unconcerned about 'spill over' effects as the smaller UDCs.

All of the contributors concur that the UDCs were, in various ways, out of touch with their localities and operated at a scale unlikely to tap local indigenous potential or respond to local pockets of need. As the chapter has already demonstrated, UDC expenditure on community projects was a small proportion of their total spend while, for Dabinett and Ramsden, an enabling and innovative urban policy should not be dependent on property cycles or be the preserve of powerful people operating at the 'flagship' scale. Likewise, for Meegan, the inability for a 'trickle-down' effect to appear in Merseyside was because of the absence of a sustained strategy in favour of MDC's 'set of ad hoc, time limited, policy experiments'. We must conclude, then, that while the operation of individual UDCs cannot be fully understood without setting them within their local contexts, in one crucial way they remained impervious to locality – namely, in their lack of recognition of, and addressing as a priority, local poverty.

Notes

1. An election pledge of the Labour Party was to integrate government departments with closely aligned functions. To this end, the Department of the Environment and Department of Transport were amalgamated into a single organisation after Labour's 1997 election victory to form a new ministry, the Department of Environment, Transport, and the Regions (DETR).
2. While the UDCs were given powers which enabled them to circumvent local democratic channels, many chose not to do so because the only way for them to get on was to attain some measure of co-operation with local actors, particularly local government. Thus, as chapters in the book will illustrate, the UDCs were, of necessity, enmeshed in local networks, with some development corporations, for example, actively forging relations with local authorities by developing joint committees and committee cycles, and entering into partnership agreements and utilising shared budgets. Indeed, for many commentators, the degree to which UDCs were able to achieve specific regeneration targets is measurable, in part, by the types of connectivity they developed with local actors and the extent to which they were willing and able to forge local partnerships.

3. The Rate Support Grant, or what is commonly referred to as the 'block grant', is the proportion of local authority revenue which is provided directly by central government. For additional details on the block grant see, for example, Wilson and Game, (1994).

4. The Urban Programme became operational in 1969 and identified 57 different areas in England and Wales as places demonstrating 'special social need' (Atkinson and Moon, 1994).

5. UDCs were also part of an emergent contracting culture in the 1980s whereby much of their preparatory, investigative, and research operations were contracted out to private sector consultancies. While this, in itself, was not an issue, some commentators queried both the scale of resources being contracted to consultancies and the tender practices by which contracts were being awarded (NAO, 1997). For instance, BDC spent £14 million of its lifetime grant-in-aid of £78.9 million on consultancy contracts, with LDC spending £5 million on consultancies from a lifetime grant-in-aid of £55.7 million. For the NAO (1997), however, the significant issue was the veil of secrecy drawn across tendering processes, and the capacity for contracts to be let on a more-or-less non-accountable basis. For instance, the NAO (1997: 46) examined 13 of the 35 consultancies used by BDC and 'found weaknesses in the employment of consultants'. In one instance, a consultant was employed by BDC to galvanise the corporation's public image. However, contrary to DoE (1990) advice, the consultancy operated in a clandestine manner without revealing that they were in the employ of the UDC.

   As the NAO (1997: 48) have stated about this case: 'a public relations company was paid to, among other duties, maintain direct contact with Ministers and their offices and to interest backbench Members of Parliament in the activities of the Corporation and to encourage them to support the Corporation. The consultant organised meetings for the Corporation with Members of Parliament, and representatives of public and private sector bodies; drafted letters to Ministers seeking support for the Corporation; and prepared a draft letter for the leader of the Conservative Group on Bristol City Council to write to the Minister of State for Housing, Inner Cities and Construction in defence of the Corpoation's activities and opposing the local authority. The consultant also wrote to the press, as a concerned citizen, complaining about an article criticising the Corporation, without disclosing that he was being paid to work on its behalf'.

6. Even if one were to accept the key output measures adopted by the UDCs, as the basis for evaluating them, there were still difficulties and inconsistencies with the ways in which such measures were adopted and used by the respective UDCs. As the NAO (1993) have observed, UDCs interpreted the key output measures differently or, as they noted:

   > some UDCs measured jobs gross, others net. In some cases outputs claimed in terms of commercial floorspace and dwellings provided included projects still under construction; but on others only the results of completed projects were included. Until 1990 therefore, when definitions were standardised . . . it was difficult to arrive at consistent measures of UDCs' performance or to make valid comparisons. (p. 11)

7. There is little evidence of a trickling-down of the economic benefits of physical regeneration to the original residential populations of UDAs and immediately adjacent areas. For example, evidence from Cardiff Bay and London Docklands is

presented in the chapters which follow which suggests that trickle-down is insignificant. Robson *et al.*'s (1994: 29) analysis of local circumstances in three conurbations, each of which contained at least one UDC, was similarly sceptical. As Robinson and Shaw (1991) point out, however, this does not mean that physical redevelopment on the UDC model is irrelevant; simply that it is not sufficient to secure lasting benefits for poor people. Nor does it mean that there were no benefits for any local people.

8. For instance, Bristol DC left £5 million worth of liabilities and 'did not comply with government guidance on land disposals' (quoted in Silke, 1997: 3). The NAO (1997) also noted that more than 100 tasks were not completed by Bristol by the time of its exit while 'a £200,000 endowment by Leeds UDC to Leeds City Council, for the maintenance of landscaped areas, was known to be insufficient' (quoted in Silke, 1997: 3). Moreover, inadequate statements, on outstanding regeneration needs in both cities, were deemed to be made by the respective UDCs (NAO, 1997).

Further reading

The best general text on urban policy is Atkinson and Moon (1994). Robson *et al.*'s (1994) heroic effort at an assessment of the impact of a rag-bag of policies with ill defined objectives, on the basis of inadequate data, still provides an invaluable source of quantitative information and systematically collected judgements of policy makers, residents in deprived areas, and businesses. Recent national urban policy initiatives are being evaluated at the time of writing with some interim statements available such as that by Russell *et al.* (1996). Readers should also have a look at Cochrane's (1993) excellent book *Whatever happened to local government?* to gain an accessible overview of the transformation in local government of which the UDC story was a part.

PART II

The British Urban Development Corporations: Policies and Practices

2

Turning the East End into the West End: the lessons and legacies of the London Docklands Development Corporation[1]

SUE BROWNILL

Introduction

> Docklands has come of age as a city in its own right . . . an equal partner with the West End and the City (Michael Pickard, last Chairman of the LDDC, *Docklands News*, April 1988)

Transforming the East End into the West End by extending the activities of the City and creating 'balanced communities' out of a predominantly working class area had been on the planning and political agenda from the time of the first Docklands plans in 1972 (Travers Morgan, 1972). By the end of the LDDC's (London Docklands Development Corporation) lifetime in March 1998, it appeared that the eight and a half square miles of Docklands had indeed gone west with 25 million sq ft (2.3m sq m) of commercial space built, a new office centre to promote London as a World City at Canary Wharf, 24,000 homes and a level of owner occupation up from 5% to 43% (see Figure 2.1). The Urban Development Corporation had proved itself to be the form of urban governance that could bring global change to the inner city. As this chapter will show, the balance between east and west is more complex than might at first sight appear – nevertheless the extent and depth of change cannot be denied.

And it is not just in the physical outputs that the landscape of the East End has been transformed. With the exit of the LDDC many of the boroughs which had previously resisted city-type developments, far from looking for a major change of direction, began working towards redressing the remaining imbalances between west and east London. By the same token the LDDC left trumpeting its outputs rather than the political strategy that had set it up. Docklands is no longer 'an ideological, noisy place' (Bob Colenutt, interview, 1988) epitomised by the clash of opposing visions for regenera-

Figure 2.1 *Canary Wharf under construction*

tion. This suggests that the map of urban governance and local politics has also been altered by the last seventeen years of regeneration activity and its political and economic context.

As the UDC experiment draws to a close it is important to look at the lessons and legacies that emerge from this remodelling of East London. This chapter therefore aims to draw out some wider issues about UDCs by first looking at the nature and extent of the changes that have resulted and raising questions about whether the locality has indeed been overridden by global change and who has benefited. Secondly, some lessons from the LDDC as a form of governance and a strategy for regeneration will be considered. And, finally, by looking at the approaches of successor agencies the chapter will explore whether these lessons have indeed been learnt or whether the legacy of the LDDC lives on in the future regeneration of East London.

A brief history of the LDDC

Before discussing these issues in more detail it is important to remind the reader of the history of Docklands' regeneration under the LDDC and what the experience of what was very much the flagship UDC can say about the lessons and legacies of the initiative as a whole. Many of these lessons and legacies have already been identified (see Brownill 1990, 1993a and b; Brownill *et al.*, 1996; Imrie and Thomas, 1993b; Newman and Thornley, 1996). These include: issues around the governance of urban policy, who is included and excluded by non-elected agencies, and the contradictory implications of the processes of what has been termed entrepreneurial governance (Du Gay, 1996) including, for example, speed and officer discretion. In terms of strategy major lessons include the gap between the rhetoric of market-led regeneration against the reality of significant public investment, the impact of booms and slumps in the property cycle, the problems associated with the lack of co-ordination of land-use and infrastructure and the failure of 'trickle-down' to ensure widespread benefits. These are illustrated in the following section and are discussed in more detail later.

In considering the balance between east and west, as the previous edition of this book indicated (Imrie and Thomas, 1993b), UDCs were not immune to local influence, despite being able to open up areas to large-scale capital investment. Further, the UDA covered six and a half square miles, parts of three boroughs and numerous smaller communities. Not only is it therefore impossible to talk about the Docklands community but also the variations between these areas meant a different balance between local and extra-local forces was achieved in each. Finally, the LDDC's approach to regeneration was not constant over the seventeen years of its existence. It is possible to distinguish four phases of LDDC activity (see Table 2.1): 1981–1985; 1985–1987; 1987–1992 and 1992–1998. These evolved as a result of a number of factors including local influence and opposition, changes in central

Table 2.1 Phases of LDDC activity

---

1. Priming the pump 1981–1985

Land acquisition and disposal	Private housing
Publicity and marketing	Docklands Light Railway
Area Frameworks	London City Airport
Targeting of high tech	Daily Telegraph and other EZ developments
Lack of consultation	Early opposition
Physical regeneration	Lack of social investment

2. The Second Wave 1985–1987

Large-scale developments	Canary Wharf
Targeting of financial services	Royal Docks schemes
Land and house price boom	Docklands Highway
Infrastructure problems begin	DLR extensions
Beginning of 'social regeneration'	Change in rhetoric
Tories win '87 election	

3. From Physical to Social Regeneration? 1987–1992

Agreements with boroughs	Community Services Division (CSD)
More consultation	Consultative structures in the Royals
Property market crash	Canary Wharf in administration
Shift in national policy	Partnership
LDDC in the dock	Critical Parliamentary reports
Financial crisis	Cost of transport

4. Preparation for Exit 1992–1998

CSD closed down	Social spending falls
Negotiations with boroughs	Exit packages
Phased de-designation	Community Development Trusts
Influence planning	Flood of applications, UDP negotiations
Property market picks up	Canary Wharf bought back, phase 2 starts
English Partnerships takes over in the Royals	LDDC writes its own obituary

---

government policy, economic trends as well as changes in LDDC's strategy and they are outlined in more detail below.

Phase one, priming the pump: 1981 to 1995

The early days of the LDDC have been covered in detail elsewhere (Brownill, 1990 and 1993a) and will therefore be only summarised here. The strategy during this first phase was to kick-start development through investment in land and infrastructure and marketing the area. Thus from the start the ideology that development was market-led was being contradicted. During this time LDDC acquired further land and initiated transport infrastructure in the form of the initial Docklands Light Railway (DLR) to run from Tower Hill and Stratford to Island Gardens on the Isle of Dogs (see Figure 2.2) and London City Airport in the Royal Docks. Attracting development and 'filling up the Isle of Dogs' took precedence over any strategic approach to economic development. The era saw, for example, the beginnings of the wholesale relocation of Fleet Street into Docklands with the *Daily Telegraph* moving its printing press to the Enterprise Zone (to be followed later by its offices).

Planning was to be swept away in preference for the promotion of catalytic or flagship developments which would determine the urban environment around them. Thus planning was turned into a marketing exercise to go alongside the rebranding of Docklands which the LDDC was undertaking to entice investment through advertising and, in particular, promoting the waterfront nature of the development. In this way the locality was turned merely into a backdrop to development. This was compounded by the belief in 'trickle-down' and therefore no attention was paid to trying to ensure that local residents gained from developments.

Governance and politics in this era was characterised by a high degree of conflict. Local boroughs, angered by the imposition of what they saw as an alien and unaccountable body, chose not to co-operate or even, in the case of Southwark, have any contact with the LDDC. Given that the LDDC considered consultation to be 'a relic of the local authority days' (quoted in

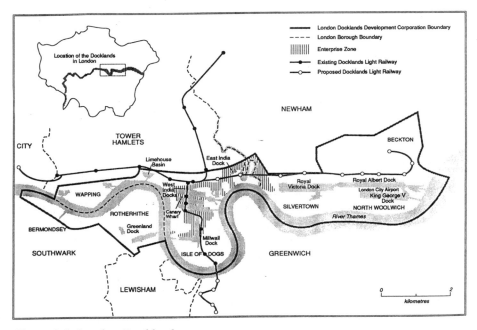

Figure 2.2 *London Docklands*

Brownill, 1990) many community organisations were similarly excluded from decision-making. LDDC's style of operation was also different, mirroring the private sector and going for quick results rather than following the bureaucratic procedures of local government.

Outputs at this time were fairly modest, as was the level of LDDC expenditure. By March 1985 the LDDC had spent £205m, 5,700 jobs were claimed to have been created with a total of 30,000 predicted, 2,466 housing units had been completed with a further 4,617 under construction and 1.9m sq ft of floorspace built (LDDC, 1985).

Phase two, the second wave: 1985–1987

All this was set to change post-1985. As the UK economy boomed so did Docklands. Property consultants Savills (1997) reported that Docklands residential prices grew by 128% between 1984 and 1988 with 73% of this growth occurring between June 1985 and June 1987. A speculative market developed in Docklands flats with pre-sold properties changing hands several times before they were built and the myth of the Docklands yuppie, who recycled City bonuses into such flats, was born. This indicates the tendency of a regeneration strategy built around property development to mirror fluctuations in the market.

The coming of Canary Wharf both was a result of and intensified this boom. The proposal totalled 10 million sq feet of office and related use and was justified as being vital for ensuring London retained its position as a world financial centre. Despite the fact that the LDDC's commercial strategy had shifted to targeting the financial services sector ahead of the deregulation of the City of London in 1986, the ex-chief executive of the LDDC called Canary Wharf a 'happy accident' rather than a result of deliberate planning (Bentley, 1997 p. 52). The ripple effects of Canary Wharf were felt not only in the increase in land and property prices but also down river where three major development proposals were put forward for the Royal Docks.

The coming of Canary Wharf and other schemes highlighted other aspects of the LDDC approach. First, the lack of planning meant the transport infrastructure in the area was totally inadequate to cope with the now projected 50,000 employees in Canary Wharf alone. This led to proposals to upgrade the railway and add an extension to the underground at Bank and for a major road system through the area called the Docklands Highway. Secondly, the precariousness of a property-led strategy was underlined by the failure of the original Canary Wharf consortium to secure financial backing for the scheme, leading to the desperate search for other backers which ended with the signing of a deal with Olympia and York in 1987.

In terms of governance and politics the ability of UDCs as a form of governance to push through major developments was underlined by the Canary Wharf scheme. As it was mostly in the Enterprise Zone it did not need planning permission and the LDDC Board did not, of course, provide a public forum for debating the scheme or its strategic impact. Close links between Thatcher and the Reichmanns (Figure 2.3) confirmed the patronage elements of UDC governance (Coulson, 1993). The Docklands Highway was also largely excluded from strategic consideration as it was presented as separate local access roads, not one major trunk road. Such events had inevitable con-

Figure 2.3 *A Thatcherite project?*

sequences for relations with the local boroughs and local communities who felt that they had been forced to accept developments with no consultation. Opposition continued and community organisations, in part funded by the GLC, were effectively using the media to expose the LDDC's failures to invest in the local community (see for example Docklands Forum, 1987). But the beginnings of a change in approach were also observable. Christopher Benson, the then Chair, talked of a 'second wave' of development whereby some of the value of developments could be diverted to fund social and community gain (LDDC, 1987). This was to be extended in later phases.

The increased pace of development was shown in the figures. By 1987 it was claimed that there were 36,300 jobs in the area as opposed to 27,000 in 1981, 9,000 residential units were built and a further 6,000 were under construction and LDDC total expenditure rose to £356m.

Phase three, the LDDC in the dock: 1987–1992

This was a time of change and crisis in all areas of LDDC activity. In terms of strategy, the initial part of this phase was marked by a concern with social regeneration. LDDC signed two agreements with local authorities, the Memorandum of Agreement in the Royals and the Tower Hamlets Accord. The latter secured rehousing for households whose flats were demolished to make way for the Docklands Highway plus £35 million in social and community projects. The Memorandum included goals for 1,500 social housing units in the area, social and community facilities and a structure for consultation. A Community Services Unit was set up within the LDDC which, in the two years 1989/90 and 1990/91, spent £112m on social housing, training, childcare and other social projects representing 18% of total LDDC spend in those years (LDDC, 1990 and 1991).

These changes were in response to a number of factors. First, Parliamentary reports were published which criticised LDDC's lack of social investment (House of Commons, 1989; NAO, 1988). Secondly, private developers were also aghast at the social mess in the area and realised before the LDDC that not only was a skilled and trained workforce important for the commercial viability of schemes but so was a neighbourhood free of social conflict and possible unrest.

Thirdly, the election of 1987 served to introduce a change in relations of governance and politics. Labour boroughs realised a change of regime was not going to happen and so shifted to a more pragmatic response, negotiating for gain and taking up seats on the LDDC Board. Some community groups followed suit. But opposition did not disappear, nor did the basic structure of governance change. The SPLASH (South Poplar and Limehouse Action for Secure Homes) campaign, around the rehousing and the environmental impacts caused by the construction of the Docklands Highway and Canary Wharf, was active. A court case taken out by the community against Canary Wharf and the LDDC is still pending.

Whether or not these changes represented a shift to social regeneration or marginal additions to an unchanged agenda is open to debate (see Brownill 1990, 1993a and b). This was, in part, due to the problems brewing at this time. The property market crash of 1987/88 hit Docklands particularly badly. The E14 postal district (which covers the Isle of Dogs) saw property prices fall by 43% between 1988 and 1994, the largest fall in London (Docklands Forum, 1994). By 1992, 45% of Docklands' office space was vacant and Canary Wharf itself was taken into administration in that year with only 60% of the space let. The property crash was heightened by the perceived problems of getting into and out of Docklands, the result of landuse and transport infrastructure being uncoordinated.

The costs of providing infrastructure were also proving to be great. LDDC expenditure between 1988 and 1992 was a massive £1.264bn, three times the total spend to 1987 (LDDC Annual Reports, 1981–1992). The Limehouse Link Road, built at a cost of £450m for 1.5km, proved to be the most expensive road in the country for which the LDDC received severe criticism from the Public Accounts Committee (National Audit Office, 1995). But this reflects the desperation of the LDDC to put in the transport infrastructure and the influence of the Reichmann Brothers of Canary Wharf over government decisions. But even their power was not sufficient to convince the government to proceed with the Jubilee Line, which was seen as vital to Docklands, without a £400m developer contribution. When Canary Wharf went into administration the chances of this being paid disappeared. LDDC was in a financial crisis as income from land sales dried up and central government sought to reduce public expenditure. As a result the social regeneration programme was heavily cut.

By the end of this period LDDC's image was one of how not to regenerate. Companies had stopped trading and developments failed to materialise, particularly in the Royal Docks. The flagship development had failed at Canary Wharf and the UDC experiment itself had been told to wind down by right wing ministers (Portillo and Redwood) who were both horrified at the levels of public expenditure and who ironically saw UDCs as an unnecessary interference in the market. Yet figures show the result of the boom: 2.02 million sq m of floorspace, 16,200 new dwellings and 63,500 jobs in the area by 1992 (LDDC, 1992).

Phase four, the end of the road for the LDDC: 1992–1998

The last phase of the LDDC's life was dominated by preparations for exit and the phased de-designation of the area. In terms of governance and politics there was a further change in approach as the LDDC realised that in order to finalise its work and hand over assets and liabilities it had to work closer with the local authorities in the area. With the replacement of Thatcherism with Majorism, the national regeneration policy context was going through some radical changes with the promotion of the partnership approach and the stressing of holistic regeneration. The slump also meant that LDDC had no major schemes to bulldoze through, opening up a space for negotiation. The boroughs themselves were more than happy to access the responsibility and budgets that such a relationship would bring. Therefore the changes in relations with the boroughs that had begun after 1987 were consolidated in the 1990s. As a sign of its more outward looking approach and the extent to which it was becoming embedded in local politics LDDC representatives also sat on regeneration partnerships in the area, including the Thames Gateway London Partnership. Yet, as we shall see, a closer relationship between the boroughs and the Corporation was not necessarily repeated at the level of the community. Nevertheless in 1997 a MORI poll undertaken for the LDDC revealed that for the first time in its lifetime more people thought LDDC took account of their views than did not (LDDC, 1997a). However, there was still localised opposition, for example, over a riverside site in Wapping where local residents wanted a memorial park to civilians killed in the blitz and the LDDC and developers wanted luxury housing.

The closer links with the locality were not necessarily reflected in strategy. The community investment budget was slashed and the Community Service Division (CSD) closed down as the LDDC was forced to get its financial house in order and sought to reduce social housing investment in its concern with balanced communities (Docklands Forum, 1996). For example, in 1994/5 LDDC spent under £7m on community support and housing, representing only 6% of its budget (LDDC, 1995). However, the recession opened up the possibilities for developments with a large public sector input, particularly in the Royals. Thus on the north side of the Royal Albert Dock a new University Campus for East London University is being developed, representing the only major public sector building supported by the LDDC. Single Regeneration Budget funds have also contributed to this development. On the side of the Victoria Dock, Britannia Urban Village has taken shape as a mixed tenure scheme with social and community facilities built in from the start. Planning for real exercises were also held during its design stage.

The decision to fund the Jubilee Line was confirmed in November 1993, secured by Canary Wharf being taken out of administration to receive a loan from the European Investment Bank covering the developer contribution. This helped lift the property market out of recession and reinstate the emphasis of private development. In 1996 Canary Wharf was bought back by a consortium, including Paul Reichmann, and, by 1998, it was 98% let with building work proceeding on new headquarters for Citibank and HSBC which will add two new towers to the Docklands skyline. In 1997 property prices on the Isle of Dogs increased by 30%, reminiscent of the boom of the late 1980s and, as we shall see, there was a flood of applications for residential and commercial use. Demand from far-eastern investors seeking investment property helped fuel these rises.

Other significant events included the IRA bombing at South Quay in 1996 which, according to the LDDC (1998a), made the LDDC into a victim and therefore helped its public image. However, it is the arrangements for exit which are most indicative of this time and which also show how the LDDC will continue to exert influence over Docklands development. De-designation was phased, starting in Bermondsey in October 1994 and ending with the Royals in 1998 (see Table 2.2). Within this there were various priorities. One was asset disposal, although, perhaps in response to problems at Bristol and Leeds (National Audit Office, 1997), LDDC did not initiate a 'fire-sale' of cheap land. The unlikelihood of selling remaining sites in the Royals was underlined by the announcement in 1995 that English Partnerships would take over land holdings there after the LDDC's departure.

A second priority was to secure the planning future of sites not already developed. This was done through influencing the boroughs' Unitary Development Plans (UDPs) which were being drawn up at this time. In Newham, for example, the LDDC originally put forward 300 objections to the UDP, although these were later negotiated away and the Corporation appeared at the inquiry in support of the compromises that formed the plan. The granting of planning applications also ensured a posthumous influence over development. In Tower Hamlets, in the month before withdrawal from Wapping and Limehouse, the LDDC received applications for over 1,100 residential units in the Tower Hamlets section of the UDA. This represented 17% of all residential completions in the area between 1981 and 1997 and cannot be wholly attributed to an upturn in the property market but rather suggests exit had focused developers' minds.

Perhaps the most interesting aspect of the exit arrangements was the negotiation with the boroughs of a package for each of the de-designation areas. These packages were basically mechanisms whereby the boroughs would take on liabilities, e.g. the maintenance of open space and roads, in return for LDDC funding projects, such as housing improvements and community centres, before its demise. In the first package Southwark received £1.5m in projects in return for the borough maintaining various pieces of open space in Bermondsey Riverside. These increased to £5.6m in Beckton and £9m in the Royals.

Included in some of these packages, notably on the Isle of Dogs and the

Table 2.2 De-disignation dates

BERMONDSEY RIVERSIDE	31 OCT 1994
BECKTON	31 DEC 1995
SURREY DOCKS	20 DEC 1996
WAPPING	31 JAN 1997
LIMEHOUSE	31 JAN 1997
ISLE OF DOGS & LEAMOUTH	10 OCT 1997
ROYAL DOCKS	31 MAR 1998

Source: LDDC (1997b)

Royals, were capital endowments to Community Development Trusts which would continue revenue funding for the voluntary sector after the LDDC had gone. Thus the Isle of Dogs Community Foundation received £1.4m and the Royal Docks Trust £2.7m. Too late the LDDC found a way of turning its capital investment into the revenue funding that community groups had long argued for.

These packages were partly essential to secure the handover of assets but they are indicative of the changed relationship with the boroughs. Other motives may have played a part such as the desire to leave with a good image and the fact that behind the scenes negotiations were going on about extending the LDDC's life by transforming it into an East London Development agency with responsibility for Thames Gateway and other sites. Therefore the LDDC would want to espouse partnership as a way of trying to convince decision-makers that it could move into the regeneration ideology of the 1990s.

Community organisations were, however, excluded from the secret negotiations between the LDDC and the boroughs (Brownill, 1998). On the Isle of Dogs and in Beckton, groups were proactive producing reports that showed the regeneration gap between local needs and what had been built over seventeen years (Association of Island Communities, AIC, 1997). However, strict Treasury guidelines meant the asset/liabilities equation could not be increased to include these costs.

Part of the exit strategy involved LDDC, in effect, writing its own obituary in order to influence opinion of its legacies. Monographs (e.g. LDDC, 1997c), a video (LDDC, 1998b) and a glossy coffee table account of the Docklands story (Bentley, 1997) all stressed the positive aspects of LDDC's activities. An example, as one community representative noted, of 'the winner writing history' (Brownill, 1998).

The final piece of the exit jigsaw was the involvement of successor bodies. As with other UDCs, outstanding liabilities were transferred to the New Town Commission and the assets of the Royals were transferred to English Partnerships. On the Isle of Dogs the dock estate has been put in the hands of the British Waterways Board for management. All of these are unaccountable bodies and, in the case of English Partnerships, likely to be short-lived as it is likely to be superseded by the Regional Development Agencies.

The lessons and legacies of the LDDC

This brief history shows some interesting changes in the LDDC's approach, relationships with the locality, and how it was viewed from the outside. It evolved from an ideological standard bearer through a phase both of crisis and a shift towards local and social concerns and emerged embedded in the locality, stressing its success in terms of outputs. These changes can be seen in the reactions to its exit. According to one Tower Hamlets officer, while some have put the flags out many are concerned at the loss of LDDC's budget particularly for marketing and for voluntary sector support. Yet of more concern are those reactions that do not seek to critically address the lessons that emerge from this history. For example, the LDDC has been held out as a model for a future London Mayor and Assembly to follow (Travers, 1988) and Regional Development Agencies look set to mirror many UDC features. A senior SRB manager in East London professed to be 'unsure of what the lessons are' (quoted in Brownill, 1998). This may be a sign of the times and how far the politics of regeneration has changed since the 1980s but it also suggests that regeneration in East London and elsewhere may end up ignoring the lessons and legacies of history. The following section attempts to draw some of these out.

Has the East End become the West End?

One of the more lasting legacies of the LDDC will inevitably be the impact of the development that it initiated. Table 2.3 confirms some of the changes that have occurred. The image and culture of the area has also been transformed both by the developments and by their marketing. The LDDC's self-evaluation (Bentley, 1997) likens the bright new shiny Docklands to a Xanadu on Thames. Can these changes really be seen, as Michael Pickard claims, as the emergence of Docklands as a second City? Inevitably there are various issues which need to be raised in relation to this. One is that the images of the City and the West End are, in effect, euphemisms for the opening up of Docklands to the demands of capital and to global economic and cultural forces (Cohen, 1996; Eade, 1998). The East/West dimension therefore becomes one of the balance achieved between local and extra-local relations. There can be no doubt that Docklands has been opened up to such processes, yet this section will show that it has not been one-way traffic.

As an indicator of this, regeneration has not been uniform across all the Docklands area. As Table 2.4 shows, there is a world of difference between the Isle of Dogs – which, for example, has seen 75% of the total office development in the area, a higher than average proportion of private housing (82%) and an occupational structure of 77% white collar jobs – and the Royal Docks where 45% of the housing development has been social housing, only 1% of total UDA office space has been built and 37% of jobs are semi- and unskilled (LDDC, 1997b and 1998b).

These differences can be attributed largely to proximity to the City, EZ incentives, and water frontage, which had an influence on the land and property markets in the areas and the influence of local politics. For example, the Memorandum with Newham council influenced the numbers of social housing units built in the Royals. Thus while Bermondsey, Wapping and the Isle of Dogs all feature in City property guides and Shad Thames in Bermondsey even outprices City property prices, residents of Beckton assert, 'Beckton used to be working class, now it's posh working class' (Brownill, 1998). Therefore commentators who see Docklands as an outpost of 'multi-

Table 2.3 Docklands before and after the LDDC

	1981	1997	Forecast
Population	39,429	81,231	112,054
Employees	27,213	72,000	175,000
Home ownership	5%	43%	
Dwelling stock	15,000	35,665	
Service sector employment	31%	70%	
Financial services employment	5%	42%	
Commercial floorspace since 1981	2.3m sq m		
Housing units since 1981	21,615		

Source: LDDC (1997b)

Table 2.4 Regeneration by area of docklands, March 1997

	OFFICE SPACE sq metres	TOTAL COMMERCIAL sq metres	HOUSING UNITS	SOCIAL HOUSING Units	% PRIVATE HOUSING
ISLE OF DOGS	1m (75)	1.4m (60)	4178 (19)	754 (13)	82
WAPPING	155,192 (11)	303,717 (13)	3874 (18)	653 (11)	84
SURREY DOCKS	169,374 (12)	370,172 (17)	7654 (35)	1843 (32)	75
ROYAL DOCKS	12,673 (1)	234,348 (10)	5909 (27)	2638 (45)	55
UDA	1.38m	2.23m	21,615	5968	72

Source: LDDC (1997b)

NB figures in brackets are percentage of total UDC development in a particular category

cultural global capitalism' (Cohen, 1998) need to be cautious about putting all Docklands under this heading.

In fact, in terms of global city status Canary Wharf is an 'island of city-type development in a sea of local commercial and residential development' (*Property Week*, 15 May 1998). But it has to be remembered that this is only 71 out of 4,000 acres and that even at Canary Wharf rent levels (at £30–35 per sq ft in Canary Wharf and £25–20 elsewhere) are still substantially below those in the City (£40–50) and West End (£55–60) suggesting it is precisely because Docklands is not the City that it is attracting tenants (*ibid.*). If global status is dependent on command and control functions it should be noted that only 3% of establishments in Docklands are headquarters offices although this is set to increase, but Docklands does now house a significant number of newspapers (LDDC, 1998b).

East meets west and global meets local in a variety of other ways. The recoding of the area's heritage, particularly dock cranes and buildings and the water itself, as a marketable characteristic to distinguish Docklands from any other office location, is one example. But of more significance is the impact of the restructuring of housing and labour markets and how this has interacted with the local social relations in Docklands.

With 42% of Docklands jobs now in the financial services sector as opposed to 5% in 1981, and a similar percentage of housing being owner-occupied, the emergence of a 'balanced' community has apparently been secured (unfortunately lack of up to date data prevents a fuller examination of this). Yet the dramatic changes in housing and labour markets raise questions about who has benefited from regeneration. The Isle of Dogs may well feature in City property guides but one in three families there are on income support and 40% of children are growing up in no-earner households (AIC, 1997). A recent report on Bermondsey and Rotherhithe notes:

> This is an area which suffers from its image. It may have been redeveloped but it has not been regenerated. Side by side with housing that is available only to the richest few, there are overlooked pockets of severe poverty and urban stress (Time and Talents, 1997, p. 3)

The fact that only 27% of the jobs in the area are new in the sense that they are not relocations (LDDC, 1998b) has led to a limited impact on local unemployment figures. For example, between 1981 and 1996 the numbers unemployed in the UDA rose from 3,533 to 4,673. Due to the increase in population this meant an actual fall in the rate from 10% to 8% (Docklands Forum, 1996). Localised unemployment amongst wards with extensive social housing remained high. For example the rate in Blackwall rose from 22% to 29% between 1981 and 1991. The existence of severe skills mismatches which further restricts local uptake shows that regeneration is a more complex process than merely building buildings and that City living is not the experience of all Docklanders.

These figures indicate the failure of trickle-down and the limited investment in the community. Table 2.5 shows that the LDDC spent only 12% of its total investment to 1997 on social and community support. Nevertheless,

Table 2.5 Summary of LDDC expenditure by area of expenditure

	£m
LAND ACQUISITION	187 (9%)
LAND RECLAMATION	157 (7%)
UTILITIES	159 (7%)
ENVIRONMENTAL	149 (7%)
ROADS AND TRANSPORT	662 (30%)
DOCKLANDS LIGHT RAILWAY	312 (14%)
SOCIAL HOUSING	163 (7%)
COMMUNITY AND INDUSTRY SUPPORT	117 (5%)
PROMOTION AND PUBLICITY	27 (1%)
ADMINISTRATION & MAINTENANCE	261 (12%)

Source: LDDC (1997b)

these levels of investment were probably higher than they would have been able to achieve given the restrictions on local government spending in the 1980s. However, as part of the exit negotiations Island community organisations commissioned a report which indicated that £9m was needed to fill the regeneration gap between the needs of many local residents and the new complexes at the end of their streets (AIC, 1997). Finally regeneration has interacted with other divisions in the locality, notably race. The election of a British National Party councillor on the Isle of Dogs indicated how the restrictions in the supply of social housing, in part a result of regeneration, exacerbated racial tension in the area.

Therefore while Docklands has been laid open to large-scale capital investment this has to be seen as the interaction between local and extra-local relations rather than a wholesale transformation.

LDDC *and the governance of urban policy*
The issue of governance continues this theme of the balance between East and West but is also an important area where lessons for future regeneration can be drawn. Docklands says much about the role of governance as the link between local and extra-local forces and influences. UDCs represented a structure of governance to exclude local influence, bring in the private sector into a form of local corporatism and lay the basis for major change (Thomas and Imrie, 1993). A striking theme in separate interviews with Docklands actors has been references to the 'colonialism', 'fiefdoms' and 'patronage' that were perceived as a result of this structure and the processes of entrepreneurial governance associated with it (see also Chapters 1 and 11 in this book). But our brief history has shown that this form of governance was not immune to local influence and the attempt to impose change without legitimation and without some form of local dialogue ultimately proved unsustainable.

Yet this influence was mediated in certain ways and was variable across policy areas. Community groups found the LDDC more open when it came to discussions about social facilities than strategic issues such as the exit agreements. Council officers and members in interviews recounted that even with the changed relationships in the 1990s there were definite limits to what the LDDC was prepared to discuss. As shown in related work (Brownill, 1998) LDDC did not evolve into a partnership in the sense that all actors had at least the opportunity of sitting round the table and although there was some redistribution of power and influence, particularly to the boroughs, this was limited. Without formal structures influence was also mediated through networks and individuals: 'if you had the ear of the chief executive you got what you wanted' recalled one community representative (Brownill, 1998). As previous research has shown (Brownill *et al.*, 1996) issues about how the lack of equalities considerations features in the structures and processes of the new urban governance need careful consideration.

The Docklands experience reveals important lessons about the contradictory aspects of this form of governance. LDDC officers claimed that at the end of its life the LDDC was easier to influence than the boroughs. Certainly the high levels of officer discretion and the budgets at their disposal meant that quicker and more positive responses were possible. Community liaison and projects officers also built up good relations with many organisations and 'went native' in the sense that, according to one community representative, 'you wouldn't know they worked for the LDDC' (Brownill, 1998). Therefore the LDDC as a form of governance did not exclude local interests to the extent that the theory of UDCs would suggest, the East End was not totally colonised by the West but nevertheless this influence has been marginalised and incorporated in a way which did not fundamentally threaten the overall strategy.

The end of ideology?

Finally, what lessons can be drawn concerning the LDDC as a strategy for regenerating the inner city? Docklands has been referred to by an ex-chairman of the Royal Institute of British Architects (RIBA) as 'three-dimensional Thatcherism' (quoted in Brownill, 1993a), the physical manifestation of a particular political ideology. The brief history has already shown both the gaps between the rhetoric and reality and the problems associated with this approach. However, it is important to underline this with a fuller consideration of the amount of public money that has underpinned the £7bn private sector investment in the area.

Table 2.6 shows that the LDDC itself has spent around £2bn which in itself represents 47% of the total UDC budget (DoE, 1997), but added to this is the £2.7bn in related transport infrastructure (LDDC, 1997c) and the

Table 2.6 Public expenditure in Docklands

	£000m
ACTUAL EXPENDITURE TO MARCH 1997	
LDDC EXPENDITURE	2,260
ROADS (NON-LDDC)	249
RAIL (NON-LDDC)	244
UNDERGROUND	2,290
TOTAL	4,983
POTENTIAL EXPENDITURE	
ROADS	592
RAIL	75
TOTAL	5,650

Source: LDDC (1997b, 1997c)

possible £1bn in EZ financial incentives (*Observer*, 9 Oct 1994). This takes the total to £6bn and this is without cost over-runs on the Jubilee Line and a further potential £668m in other transport schemes. With total transport investment (including £800m from the private sector) in East London of £5.3bn (LDDC, 1997c) Docklands, can be seen as much as an example of reverse leverage, where the private sector secured public underwriting of its investment, than the pump-priming envisaged in the early days of the LDDC.

The cost to the public purse of LDDC's outputs has, therefore, to be put into some perspective. Taking one indicator, cost per job, illustrates this. Even using the government's own figures, which show only grant-in-aid and do not distinguish between new and other jobs (DOE, 1997), LDDC works out at £26,185 per job, the most expensive UDC per job apart from Plymouth. Leeds managed £6,143 per job and Merseyside £21,151. If we take account of the fact that not all the jobs are new, and the additional expenditure, the figures soar to £75,655 for LDDC expenditure alone, £238,000 per job with transport investment and £255,000 per job with the tax breaks. This obviously leaves out any wider impact of the expenditure but it does indicate why the Treasury was so keen to disband UDCs.

This vast public expenditure is perhaps one reason why the LDDC has left stressing outputs rather than ideology. According to one LDDC officer there was no firework display on the final day for fear of the headlines 'Docklands money goes up in smoke'. But there are also other ways of looking at this. One is that 'the government has won' (interview with Bob Colenutt) and current policies with their stress on competitiveness, outputs and partnerships involving the private sector have absorbed this ideology albeit in a modified form. The next section looks briefly at post-LDDC regeneration in East London to see if such a consensus has emerged and, if so, whether it has enabled or prevented the lessons and legacies of the LDDC being learnt.

After the LDDC: where to now for Docklands and East London?

With the winding down of the LDDC will a different balance between 'east' and 'west' result? East London is certainly not short of regeneration initiatives. They include the Millennium Dome, the Channel Tunnel terminal at Stratford, the Thames Gateway Partnership, the Greenwich Waterfront Partnership, numerous City Challenge and SRB schemes as well as the continued developments in Docklands and the borough strategies.

What emerges from these initiatives is a sense of East London at a crossroads. The legacy of the LDDC lives on in boroughs that are 'more catholic than the pope' (LDDC, 1998a). For example, in the drawing up of a borough-wide Regeneration Strategy, Newham Council is, according to its Leader, looking for the creation of a 'place of opportunity rather than poverty and neglect' (*Financial Times*, 19 March 1998). This means, for example, recreating the image of the borough and improving services so higher income residents move in or stay rather than bidding for funds on the basis of being Britain's most deprived area. The ripple effects of the LDDC's concern with 'balanced communities' is clear and, according to the ex-chief executive of Newham, a direct result of adopting aspects of the LDDC's approach (Drew Stephenson, interview).

The London Planning Advisory Committee (LPAC) has similarly argued the need for redressing the east-west imbalance in London under the banner of securing London as a world city. 'London's continuing success as a world city requires the effective use of these resources' writes LPAC's Chief Planner (Simmons and Warren, 1998, p. 75). Within this overall strategy, in an echo of flagship development, the 'East London Development Focus' (see Figure 2.4) places emphasis on key sites of 'critical symbolic significance' and 'strate-

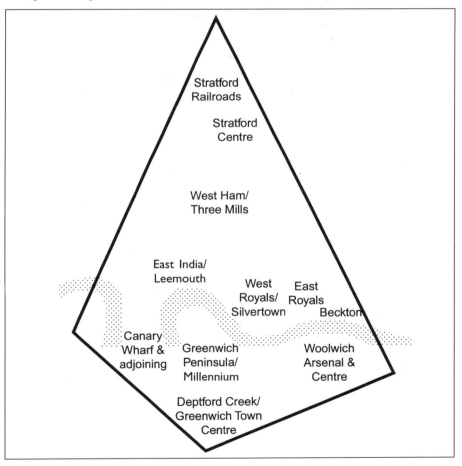

Figure 2.4 *LPAC's East London Development Focus*

gic catalysts' (*ibid.*, p. 85) in East London including the Greenwich Waterfront, the Royals, Canary Wharf and the Stratford Terminal.

But there are also signs that a different balance between east and west is being sought. The Thames Gateway London Partnership (TGLP), which evolved out of the government-led East Thames Corridor, echoes the arguments that 'East London is an ideal World City location' and relies heavily on marketing the area as such. However, there are also strategic objectives relating to local employment and securing community benefits and an economic strategy rather than a property-led free-for-all (TGLP, undated) 'closing the gap between East and West London' does not necessarily mean turning the East End into the West End.

This idea of an inclusive vision for 'World City' regeneration is taken further by the 'Rich Mix' proposal which is part of the Cityside SRB in the Spitalfields area of Tower Hamlets, home to a large and mainly impoverished Bangladeshi population. Rich Mix is a proposal which aims to underline the link between London's role as a world city and its multicultural character (Eade, 1998, p. 67). The proposal is for a building which would house an archive celebrating London's multicultural diversity plus exhibition and other space. Linked to this is an alliance between the council, Asian businesses and community organisations to establish Brick Lane as a 'Banglatown' whereby the culture and economy of the Asian community is harnessed as part of an overall regeneration strategy. This initiative may well fall victim to the pressures of commercial development and to the difficulties in sustaining cross-cultural alliances. Nevertheless it suggests that a more creative balance between east and west can be struck.

The governance of urban policy has shifted to partnerships and it is interesting to assess whether or not changing the structures will lead to a different balance between local and extra-local forces. Evidence from Docklands and elsewhere (Brownill, 1998; MacFarlane, 1993; Skelcher *et al.*, 1996) suggests that there is potential for this given the fact that local authorities are usually in the lead and community representatives are sitting round the table. But there are also constraints on local influence given the strict monitoring of central government and the reliance on extra-local sources of finance. Community involvement has also shown to be circumscribed by the processes of governance which retain many UDC-like features of an emphasis on speed, meeting outputs, financial monitoring and networks.

'We were set up as a reaction against the way LDDC was set up ... our partnership was intended to be a better way of doing things' one partnership manager said in interview. And yet, while it is undoubtedly true that the structures of these agencies have facilitated much greater involvement, networks and forms of patronage still play a major role in influencing such agencies. Indeed strategies may well contain more legacies of the LDDC than managers care to admit.

Therefore, by way of conclusion, we can imagine different scenarios for the regeneration of East London post-LDDC. One is the continuation of the trends of the LDDC era, where high value developments transform the wider East London area to the exclusion of local needs and voices. Another is the 'phoenix from the ashes' whereby innovative partnerships and key developments ensure East London becomes a global player yet the wide diffusion of benefit is achieved and at the same time regeneration is sensitive to local diversity and needs (Cohen, 1988). The balance between East and West is, therefore, redressed on the East's terms and builds upwards from its pre-existing foundations rather than being imported wholesale from the West or even further afield.

Which destination is eventually reached in East London will depend in part on whether the lessons and legacies of the LDDC era have been learnt and applied. Yet, as we have seen, the present-day consensus and output driven regeneration culture may be preventing this in full. Watching the space of East London in the future will therefore, as ever, reveal much about trends in regeneration on a national level and the continuing legacies of the UDC experiment.

Acknowledgements

The author would like to acknowledge the financial support of the University of East London and Oxford Brookes University for research which has contributed to this chapter and to thank Konnie Razzaque, Ben Kochan and Simona Florio for carrying out the interviews.

Note

1. This chapter is based on ongoing research into the redevelopment of Docklands. To date this research has consisted of thirty interviews with key actors. Some quotes are included in the text from these interviews although not all are directly attributed on the wishes of the interviewees. Brownill (1998) provides further information on the research and fuller discussion of some of the issues raised in the interviews.

Further reading

Docklands provides one of the most interesting case studies of recent urban regneration and a microcosm of the issues, debates and conflicts surrounding it. General overviews include Brownill (1990 and 1993b); Ogden (1992) and chapters in Newman and Thornley (1996). The LDDC's side of the story is published in Bentley (1997). Architectural and design issues can be explored in Edwards (1992) and Williams (1998). Post-LDDC regeneration initiatives are well charted in Rustin (1996) and in the journal *Rising in the East*.

3

Urban Development Corporations, urban entrepreneurialism and locality: the Merseyside Development Corporation

RICHARD MEEGAN

Introduction

At the time of writing, the Museum of Liverpool Life is hosting an exhibition of the history of the Merseyside Development Corporation (MDC). The Museum is itself housed, rather fittingly, in a building refurbished as part of the Corporation's extensive reclamation and restoration of the formerly derelict Albert Dock in Liverpool's southern dock system. The tone of the booklet accompanying the exhibition, and sharing its title – *All Our Tomorrows* – is decidedly quixotic, as epitomised by the introductory statement of the Corporation's Chairman, Sir Desmond Pitcher:

> Ever since its inception in 1981, every action taken by the Merseyside Development Corporation, every decision made and every penny spent has been focused on the future. Our Mission Statement has been the permanent regeneration of Merseyside's Waterfront – not just for the 80's, not just for the 90's, but for all our tomorrows. (Merseyside Development Corporation, 1997a, p. 5)

Underpinning this confident statement is the Corporation's evident satisfaction not only at having achieved, some four years earlier, all the 'standard output' targets set by the Department of the Environment when the Corporation's area was extended in 1988, but also at being well on line for achieving the more ambitious goals that the Corporation set itself when it reviewed its Corporate Strategy in 1992. Table 3.1 gives the figures for the six standard outputs. These figures are taken from the Corporation's *Completion, Exit and Succession Plan* (Merseyside Development Corporation, 1997b) which also provides the much broader list of achievements set out in Table 3.2.

The list in Table 3.2 is particularly revealing in terms of the categories that it introduces into consideration. Under the 'environment' category are

Table 3.1 Merseyside Development Corporation: achievement of 'standard outputs'

Key Achievements	Achievements to 1996	1996/97			Outturn 1981–97	Forecast 1981–98
		Target 1996/97	Outturn 1996/97	%		
Land Reclaimed (Ha)	363.1	9	18.9	210	382	388
Roads Built/Upgraded (Km)	84	12	13	108	97	100
Commercial Development ('000m²)	555	65	34	52	589	698
Housing (units)	2,875	248	260	105	3,135	3,621
Jobs created (No.)	16,595	2,472	2,510	107	19,105	22,254
Private Sector Investment (£m)	461	84	87	104	548	662

Source: Merseyside Development Corporation (1997b)

Table 3.2 Merseyside Development Corporation: widening achievements

Measure	Unit	Total
Private Sector Investment		
Direct PSI (outturn prices)	£m	548.26
Complementary PSI (outturn prices)	£m	110.68
Total PSI	£m	658.94
Employment		
Total jobs created (gross)	no.	19,105
Total jobs created (net)	no.	
Construction jobs (gross)	no.	5,600
Inward Investment		
Direct Inward Investment Conversions	no.	8
of which:		
jobs created	no.	770
direct PSI	£m	21.6
Indirect Inward Investing Firms	no.	29
Infrastructure		
Roads built/upgraded	km	97.22
Traffic calming schemes	no.	2
Public Walkways constructed/landscaped	km	12.62
Environment		
Land reclaimed	ha	381.97
of which for –		
Open Space	ha	69.9
Commercial development	ha	164.3
Housing/recreational use	ha	83.98
Waterspace restored	ha	39.42
Corridor planting schemes	ha	7.0
Canal Corridor improved	km	2.5
Contaminated dock silt removed	millions of m³	2.223

Table 3.2 *continued*

Measure	Unit	Total
Quay walls repaired	km	15.71
Historic buildings restored		672
of which		
Grade 1	no.	87
Grade 2	no.	397
Other	no.	188
Conservation areas enhanced	no.	2
Waste uses relocated	no.	8
Waste uses closed	no.	26
Designated waste management sites provided	ha	10.3
Social/Community		
Trainees trained (non-vocational)	no.	5,012
Training weeks (non-vocational)	no.	66
Festivals/events supported	no.	28
Community groups assisted	no.	138
Community space created	m^2	3,926
Training		
Training schemes supported	no.	52
Trainees trained	no.	3,020
Training weeks	no.	32,825
Unemployed obtaining qualifications	no.	456
Trained people obtaining jobs	no.	136
Business Support		
Businesses in area (1997)	no.	1,741
Firms assisted with grant-aid – IUAA, SIA, City Grant – providing:		
Private Sector Investment	£m	73.81
Jobs	no.	3,710
Business Associations supported	no.	3
Business Links supported	no.	2
Managed workspaces assisted	no.	389
Tourism		
Visits to events/attractions	no.	61.62m
of which:		
Albert Dock	no.	54.6m
IGF (1984)	no.	3.4m
Tall Ships (1984)	no.	1.0m
Tall Ships (1992)	no.	2.5m
Battle of the Atlantic (1993)	no.	0.34m
Estimated spend generated	£m	327.38
Increase in hotel accommodation	no. rms	321
of which:		
4*	no. rms	174
Budget	no. rms	147
Tourism agencies supported	no.	1

Table 3.2 *continued*

Measure	Unit	Total
Housing/Population		
Housing units	no.	3,135
Owner occupied/private rented dwellings completed	no.	1,833
Housing association dwellings completed	no.	920
Local authority dwellings improved	no.	499
Local authority dwellings demolished	no.	146
Population (1997)	no.	10,692
Commercial Development		
Commercial Development	'000m^2	589.23
Industrial development completed	'000m^2	247.60
Office development completed	'000m^2	215.05
Retail development completed	'000m^2	38.99
Leisure development completed	'000m^2	34.37
Hotel development completed	'000m^2	10.44
Other development completed	'000m^2	42.78

Source: Merseyside Development Corporation (1997b)

clear indicators of the sheer scale of environmental dereliction with which the Corporation was faced. In addition to the 382 hectares of derelict land, environmental works included the restoration of 39 hectares of waterspace, the removal of over 2.2 million cubic metres of contaminated dock silt and the restoration of 672 historic buildings of which 13% and 59%, respectively, were listed Grade 1 and Grade 2. 'Social and community outputs' are included which range across non-vocational training, the support of events and festivals as well as community groups. Measures of 'training and business support' achievements are listed alongside tourism analysed by estimated visitor numbers and spending by attraction. All of these different measures are indicative of the degree to which the MDC in its final years was a very different institution – in strategic approach, scope of activity and method of working – from the one for which the National Audit Office felt able to make a prima facie case for closure seven years into its initially scheduled ten years of operation (National Audit Office, 1988; see also Hayes, 1987; Parkinson and Evans, 1988, 1989; Dawson and Parkinson, 1990). Before the National Audit Office report was published, however, the government saw fit not only to guarantee further funding for the MDC but also to expand significantly its boundaries (see Figure 3.1). Since this second lease of life, the Corporation has certainly garnered relatively more favourable public reviews and particularly so in relation to its pioneering contemporary in London (see, for example, Dalby, 1990 and 1992; *The Times*, 1992; *Financial Times*, 1994 and 1997; *Liverpool Echo*, 1996).

In this chapter, I want to explore the history of the MDC through a perspective informed by two contemporary debates in the social sciences over, first, the significance of local, spatial variation in the operation of socio-spa-

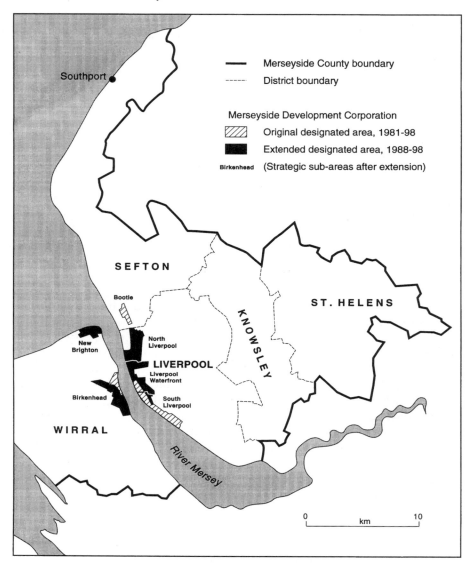

Figure 3.1 *Merseyside Development Corporation area and local authority districts, 1981–98*

tial economic processes (often referred, albeit controversially, in shorthand as the 'locality debate') and, secondly, the alleged shift in urban governance in Advanced Capitalist Countries from 'managerialism' to 'entrepreneurialism'. These debates will be introduced in the next section to provide a lens through which to view the Corporation's history. The chapter concludes with a discussion of the broader lessons that can be drawn from this history for urban and regional policy.

'Locality' and 'urban entrepreneurialism'

The 'locality debate'

There has been a resurgence of interest in the social sciences in the spatiality of socio-economic, political and cultural development (see, for example, Gregory and Urry, 1985 and Agnew and Duncan, 1989). In human geography an inspirational force has been Massey's (1995) work on spatial divisions of labour with its focus on the socio-spatial constitution of social relations. Central to the approach is the notion of waves or rounds of investment in the social and economic landscape which recondition both the physical and social aspects of place. The approach recognises that, while places are interdependent, they are also unique, representing at any given time a synthesis of political, cultural, social and economic histories and characteristics. These locally-based, historically-produced characteristics have a dialectical relationship with wider social processes. 'Geography' does not simply reflect social relationships but actively helps to mould them: space is a social construct but social relations are themselves constructed over space (Massey, 1985).

The 'spatial divisions of labour' approach informed a major ESRC-funded research initiative, 'The Changing Urban and Regional System (or CURS) Research Programme' which provoked a wide-ranging, at times exasperating and, at others, stimulating debate. Initial criticism was marked by a degree of misunderstanding (with for example, Smith (1987) – as Cooke (1987) points out – confusing the empirical research of CURS with an empiricist approach). The debate, at least in its early stages, has also been characterised at times by a rather unconstructive degree of acrimony (see, for example, the exchange between Duncan and Savage (1989) and Cooke (1989) and between Harvey (1987) and just about everyone else; and, for a calming antidote, Walker (1989) and Duncan and Savage (1991).

Central to the debate has been the concern of some that the 'rediscovery of place' and the 'locale' is a retrograde theoretical step mistakenly endowing places with causal powers. Harvey (1989a, 1989b and 1993) provides the most powerful arguments to support this concern, especially in his characterisation of the process of 'time-space compression', driven by the ever accelerating process of capital accumulation (and, in particular, the speeding-up of the turnover time of capital) that he sees at the heart of the 'condition of postmodernity'. The uncertainties created by the transformations in economic, political, social and cultural life that are associated with time-space compression, he argues, help to explain the intensified search for secure 'moorings' in 'place' and 'place identity'. It is a powerful argument but, as Massey (1992, 1993a and 1993b) has persuasively argued, it is a heavily economistic one and needs extending by both a specification of the 'power geometry' of time-space compression (because different social groups are configured differently in relation to the flows and interconnections that the process involves) and a recognition that places do not have single essential identities but multiple ones (which are not themselves constructed from an inward-looking history). Social relations for Massey are central to any understanding of 'places':

> Social relations always have a spatial form and spatial content. They exist necessarily, both in space (i.e. in a locational relation to other social phenomena) and across space. And it is in the vast complexity of the interlocking and articulating nets of social relations which is social space . . . A 'place' is formed out of the particular set of social relations which interact at a particular location. And the singularity of any individual place is formed in part out of the specificity of the interactions which occur at that location (nowhere else does this precise mixture occur) and in part out of the fact that the meeting of those social relations at that location (their partly happenstance juxtaposition) will in turn produce new social effects. (Massey 1994; p. 168)

This is an important contribution to the debate for it helps us to understand that places are expressions of both social and spatial relations, being formed out of particular sets of social relations – in the spheres of production, the state and civil society – that intermesh and interact at particular locations; and, importantly, that the social relations which constitute places are not all confined to them but are constructed and operate beyond them, connecting places and the people living in them to each other (Meegan, 1995).

Pickvance, in his (1998) response to Cox's (1998) discussion of the links between local social structures and the appeals to 'community' in local economic development, makes a particularly helpful contribution. As he reminds us, much of the locality debate is about trying to distinguish between structures and processes which are 'in' localities and those which are 'of' them. He cites the early work of such North American sociologists as Stein, who argued that the autonomy of communities was being reduced by urbanisation, industrialisation and bureaucratisation, and Warren, who distinguished between the 'vertical' pattern of relations linking localities (and units within them) to the wider society and the 'horizontal' pattern linking them to one another. But, he is also careful to emphasise that:

> neither of these authors argued that localities could be understood entirely in terms of wider processes. Rather, they insisted that localities had distinctive effects. They had histories that would affect the functioning of local structures connected to the wider society, these local structures could influence the external organizations to which they were connected; and the particular local structures present would shape each other and interact in novel ways. In my view [and one shared by the present author] all these effects capture aspects of what it is for structures and processes to be 'of' the locality, rather than simply 'in' it. (Pickvance, 1998, p. 45)

Looking at places from such a 'locality perspective' can help in understanding the spatial dimension of policy. More specifically in terms of the subject of this book, it is possible to view a particular political and policy instrument – in this case, the Urban Development Corporation – as being introduced into different localities or 'local worlds', bringing with them their own relationships (most notably with the central state) and creating their own relationships in situ; and by so doing providing new links between the places in which they are located and national (and global) processes. National policies operating across places provide, in a system sense, a skein of 'feedback links' as the policies are mediated through the particular socio-economic, political and cultural characteristics of place. Given that these characteristics are historically contingent and vary between places, policies can potentially operate with very different effects in different areas. Indeed the interactive relationships may not just produce different impacts but may also act to modify the goals and operation of the policies themselves. It would not seem unreasonable, therefore, to expect the operations of Development Corporations to differ from place to place.

The shift from 'urban managerialism' to 'urban entrepreneurialism'

Some of the key themes in the 'locality debate' – especially the links between the local and global, place identity and representation – also recur in another, equally contested, debate about the supposed evolution in the form of urban governance from 'urban managerialism' to 'urban entrepreneurialism'. This evolution has supposedly seen urban governance shift away from a 'managerialist' and localised provision of welfare and services to an outward-looking 'entrepreneurialism' emphasising growth and local economic development (for a recent review of the debate, see Hall and Hubbard, 1996). Harvey has again been a key figure in this debate, linking the transformation to a paradigmatic shift in capital accumulation from the rigid mass production and Keynesian regulation of 'Fordism' to the flexible accumulation of 'post-Fordism' (Harvey, 1989b and 1989c). It is not necessary, however, to accept uncritically the idea of the post-Fordist transformation, to agree that there has indeed been a pronounced shift to a more entrepreneurial form of urban governance – not least in the UK context where Development Corporations have played a leading role (Parkinson, 1989; Pacione, 1990 and 1997; Lawless, 1991; Thornley, 1991; Deakin and Edwards, 1993).

Harvey (1989c) usefully identifies four key competitive strategies pursued by the coalitions engaged in urban entrepreneurialism involving competition over: the international division of labour (attracting mobile investment and employment); the spatial division of consumption (tourism and consumerism – 'attracting the consumer dollar'); the acquisition of control and command functions (securing high-status activities – corporate headquarters, media and financial decision makers – in the hierarchical division of labour); and the redistribution of surpluses by central Government (transfer payments, defence expenditure etc.). All of these strategies can be detected in the operations of the Urban Development Corporations and the MDC is no exception. However, what is interesting is the way in which these strategies have been shaped by differing local circumstances and differing social relations in place.

The Merseyside Development Corporation through the lens of 'locality' and 'urban entrepreneurialism'

The Merseyside Development Corporation in historical perspective

It is possible to identify, on the basis of changing emphases in the Corporation's own strategies and policies, three phases in the history of the MDC. These phases can themselves be linked, in turn, to distinctive conjunctures in which local economic and political circumstances were articulated with broader national and global ones. Across these three periods, it is possible to trace both a gradual reorientation of the MDC's regeneration activities as local, national and international economic and political conditions change and a gradual repositioning of the Corporation in the locality as new institutional relationships and dynamics emerge. The first period, between the establishment of the MDC in 1981 and the major extension of its boundaries in 1988, saw the gradual accommodation of the Corporation's Initial Development Strategy to the harsh economic and political realities of its area of operation. The second period was a short one of transition between 1988 and 1991. It saw the Corporation attempting to come to terms with both the extension of its area and the inclusion within its boundaries for the first time of a substantial residential and working population. The third period, from 1992 to the Corporation's demise in 1998, was one conditioned by changes in national urban policy and in the Corporation's modus operandi (to a kind of mixed 'entrepreneurial and partnership' approach) and by the intrusion of a new, European, level of governance – in the form of the spending programme agreed for Objective One funding from the European Union's Structural Funds.

The re-orientation of the Initial Development Strategy, 1981–1988

The designated area of the MDC was tightly defined incorporating about 865 acres in Liverpool's South Docks and Riverside areas, parts of Bootle in Sefton Metropolitan Borough Council and a stretch of disused industrial and dock areas on the Wirral peninsula (see Figure 3.1). About a quarter of the area comprised heavily silted and polluted docks and 92% was in public ownership (with the Mersey Docks and Harbour Company and British Rail accounting, respectively, for 75 and 10% of the area and the three local authorities accounting for the remainder). In the rather despairing words of the MDC's *Initial Development Strategy*:

> Most of the area is severely degraded, being non-operational docks and back-up land, demolished goods yards and sidings, part-cleared tank farms and petroleum stores, or is land in the process of reclamation by land fill using commercial and domestic waste. The overall impression is of severe degradation, inaccessibility, danger to the public and much vandalism. (Merseyside Development Corporation, 1981, p. 5)

Looking back at the introduction of the MDC onto the local economic and political stage, a number of clear 'locality effects' can be identified which helped to shape the early history of the Corporation's activities. Dominating the complex social, economic, political and cultural make-up of the area into which the MDC was inserted was the devastating scale and pace of local economic decline.

1966 marked the peak of post-war employment growth on Merseyside. Between then and the establishment of the MDC in 1981, something like 183,000 jobs were lost in the county as a whole, a decline of one quarter. Particularly significant was the fact that 72% of the job loss was in manufacturing, the sector that had been targeted by politicians and planners as the saviour of the local economy and necessary compensation for the long-term decline of the port. A milestone in this latter decline (and to be a significant factor in the operations of the future MDC) was the closure in 1972 of the three-mile dock system south of the Pier Head as the then Merseyside Docks and Harbour Board finally caved in to mounting financial difficulties and embarked on a central government-assisted restructuring.

Unemployment returned to the political agenda with a vengeance. More than 121,000 people were registered as unemployed in Merseyside County in June 1981 representing an unemployment rate of 16.9%, over one and a half times the national rate. Population decline had begun in the 1930s and been further encouraged by the decentralisation policies of the 1950s and 1960s, but by the end of the 1970s emigration in search of work took a hold that has still to be loosened (with population declining by about 9% in each of the two decades since 1971 and distinguishing Merseyside County as the fastest declining conurbation in England and Wales).

This highly pressurised period of social and economic decline (with a particularly intense phase in the 1978–1981 recession) was finally punctuated by the outbreak of rioting in 1981 giving the area the further unwelcome distinction of containing the first city in mainland Britain to experience the use by police of CS gas to quell civil disturbance.

Social and economic degeneration was accompanied by serious environmental degradation with economic restructuring leaving a legacy of industrial dereliction and environmental despoliation. This environmental degradation was particularly severe in the core of the county and, as already noted, especially so in the sites along the riverside previously housing chemical works and refuse disposal activities and in the non-operational docks where the river system had been allowed to re-establish itself, leaving deposits of silt and mud (sometimes topped up by sewage) nearly 30 feet deep in places and overlooked by rows of warehouses in various stages of dilapidation. Eighty per cent of the land initially designated for the MDC was derelict and unused, a significantly higher proportion than was the case in the LDDC (where the figure was 45%) or in all of the subsequently created Development Corporations (see Dalby, 1990).

The MDC made much of this environmental legacy at the time with, for example, John Ritchie, the then Chief Executive, using it to fend off criticisms of the Corporation's relatively unfavourable balance between public and private investment (with the former far exceeding the latter) at the end of the first five years of its operations. He pointed to the extraordinary scale of the reclamation work that it had faced and stressed that over 50% of its public investment thus far had gone 'straight into the ground' (*Financial Times*, 1986). This theme has been frequently rehearsed by other Development Corporations, perhaps most bluntly of late by the Chairman of the Black Country Development Corporation, Sir William Francis, who has argued that 'Developers need the abnormal cost removed before they will build' (quoted in Dalby, 1990). The nature of these 'abnormal costs' is something to which we will return later in the chapter.

The goal of levering private investment (after successive rounds of private disinvestment) and of employment generation as set out in the MDC's *Initial Development Strategy* (MDC, 1981) was therefore clearly going to be severely tested in such an economically and environmentally distressed locality. And sure enough, little private investment was forthcoming, most significantly for industrial development. One indication of this lack of interest was the rental values of one of the key industrial sites in the Corporation's portfolio, the Brunswick dock area to the south of the Albert Docks. Before renovation in the mid-1980 it had rents of just 15 pence per square foot. By the late 1980s, after renovation, these rental figures had indeed increased but only to £1.50 per square foot. It was not until the 1990s that rents reached £5.00 (Bates, 1990; Regan, 1990). Strategy had to be rethought and the resultant shifts in policy were quickly reflected in changes to the original planned land uses. Key sites zoned for industrial or mixed housing and industrial development were re-zoned for housing or retail land uses (Hayes, 1987) as the MDC gradually adopted the property-led development style that typified the activities of the Urban Development Corporations in their early days (CLES, 1990b). The unwillingness of the private sector to invest in the MDC schemes, however, was not simply a reflection of the depressed state of the local economy and/or the scale of what it perceived as 'abnormal costs'. Local politics – that other important ingredient in the making of 'locality' – has also played a key role.

On balance, the MDC was given a relatively free ride in its first years from local politicians. The three local authorities directly affected by the MDC (Liverpool, Sefton and Knowsley) made no formal objections to the Corporation's establishment. The most sustained objections came from the Merseyside County Council and these were quickly undercut (first, subtly, by the appointment of its leader, Sir Kenneth Thompson, as the first Vice Chairman of the Corporation and then, more crudely, with its abolition in 1986).

The debate in Parliament of the Bill establishing the MDC clearly reveals – and for this reader at least, with hindsight, rather surprisingly – the enthusiasm of local MPs for the proposed Corporation. This enthusiasm covered the political spectrum encompassing not just the lone Tory (Anthony Steen) or the right-wing Labour MPs that were later to defect to the SDP (James Dunn, Richard Crawshaw and Eric Ogden) but also the two redoubtable left-wingers (the late Eric Heffer and Alan Roberts). For the Labour MPs, a particular welcome feature of the proposed Corporation (and again somewhat disconcertingly given subsequent furore over Development Corporations' lack of local political accountability) was its bypassing of Local Authority powers. Alan Roberts was particularly scathing about the 'Scrooge-like' Sefton District Council in his constituency (Hansard, 1981). At the time, of course, the Local Authorities were controlled by either Tory or Liberal-Tory administrations.

The MDC, unlike its London counterpart, was thus initially received with relatively little local political opposition. The situation was changed somewhat in 1983 with the election of a radical, Labour administration in Liverpool, an administration effectively led by members (or 'supporters' to use the preferred terminology) of the far-left 'Militant Tendency' (for an interesting discussion of the influence of 'Militant' in the Labour Party both nationally and in Liverpool, see Crick, 1986). The City Council was certainly critical of the MDC (see for example the comments of Tony Byrnes, Chairman of the powerful Finance Committee in the *Financial Times*, 1986). This critical stance, however, had little direct impact on the Corporation's activities as the Council became increasingly embroiled in other political battles both with Central Government over rate setting and policy direction and eventually with sections of the local community (Meegan, 1990). With their energies focused elsewhere, the Council's politicians simply ignored the MDC. Between 1983 and 1989 no-one from Liverpool City Council took up the available seat (albeit as a representative in a personal capacity) on the MDC Board (although officers did maintain the working links established by the already agreed Code of Consultation).

What did reverberate on the MDC's early activities, however, were the more indirect effects of the City Council's policies. The Council's 'municipal socialism', resolutely based on an 'Urban Regeneration Strategy' involving municipal house building and environmental and leisure development, not only jarred with the public-private approach of the MDC but also actively undermined it by alienating potential private sector investment (see, for example, *Financial Times*, 1986 and Parkinson, 1990). MDC policy thus shifted even further towards public sector-led infrastructural work and development activities for which some private sector involvement was forthcoming such as tourism and leisure and building on the success of the International Garden Festival and Tall Ships Race (both hosted in 1984) in attracting visitors to the city. The steadily growing numbers of visitors to the refurbished Albert Dock complex further reinforced this policy shift. Although even here private sector involvement was far from secure. Thus, for example, the private developer to which the MDC had handed over the Garden Festival site (against the advice of both the DoE and Liverpool City

Council and for which it attracted particularly severe criticism from the National Audit Office (1988)) quickly went bankrupt. So too did one of the developers of the first housing units on the Garden Festival site leaving the scheme in question incomplete.

Housing was not a significant feature of the first phase of the Corporation's history. Its only real residential population was confined to one block of council flats in north Liverpool, housing just 310 people, and one threatened at the time by the planned expansion of the dock estate of the Mersey Docks and Harbour Company. The Corporation provided grants towards an improvement scheme developed by the local council and the Manpower Services Commission for these flats. It also released some nine acres of the Garden Festival site for development by housing cooperatives in 1984 and began the preparation of land in the Wirral dockland area (Rose Brae) for private housing development. Planning permission was given for the conversion of former dock warehouses into apartments (with the first phase of apartments in the Wapping Dock coming on the market in 1987 and the adjacent Albert Dock the following year) and the Marina that formed part of the plans for the reclamation of the docks to the south of Albert and Wapping docks was also targeted as a site for private housing development. Taken together, the housing developments in this period were relatively small scale, however. The main housing developments were to come with extension of the area in 1988.

The official opening of the Albert Dock in May 1988 effectively marked the end of the first phase of the MDC's activities, a phase heavily dominated by land reclamation, building refurbishment and general environmental improvement. By the end of the 1987/1988 financial year, the Corporation had spent £171 million, of which 82% had been accounted for by capital works and land acquisition – with, it has to be acknowledged, some impressive results. Of the original 300 hectares of derelict land and water in its designated area, only 35 remained. What had been one of Europe's largest land reclamation programmes had been carried out in an impressively short period of time.

The perceived success of the Corporation's key 'flagship project' of the reclaimed and refurbished Albert Dock and the 'flagship events' of the International Garden Festival and the visit of the Tall Ships in attracting visitors to the area had clearly reinforced the shift in policy towards the promotion of leisure and tourist developments, and also, the beginnings of residential development on the waterfront. This policy emphasis also conditioned the form of 'urban entrepreneurialism' being adopted with promotional, 'place marketing' largely aimed at attracting visitor interest in the waterfront developments. Full-blown 'urban entrepreneurialism' in the global market place for inward investment was not on the agenda in this period (its second Chief Executive, John Ritchie (1985–1991) felt it was too expensive; see *Financial Times*, 1986). Indeed, the emphasis on land reclamation and physical works had meant that there had been no programme for business development even for indigenous firms in the first four years of operations. It was not until the end of 1985 and beginning of 1986 that a Business Development Team and Business Financial Advisor were set up and not until 1987 that a survey of businesses within the designated area was undertaken. It is not surprising, therefore, that only 2% of the MDC's expenditure of £171 million had been accounted for by support to business and training.

A combination of there being relatively little land available for disposal and a depressed land market also meant that receipts over the period were negligible. Private sector investment was equally weak. By the end of the 1987/88 financial year, the Corporation's expenditure of £171 million had only attracted some £26.5 million of direct private sector investment: one pound of public investment attracting just 15 pence of its private sector counterpart (figures calculated from MDC Annual Reports).

The MDC in transition, 1988–1991

Before the critical review of the MDC by the National Audit Office was published, the government extended the Corporation's boundaries in November 1988 (from 350 to 960 hectares, an almost three-fold increase in size) and guaranteed further funding. The Corporation now took in additional industrial and commercial areas in South Liverpool, parts of Liverpool city centre near the Liverpool waterfront, parts of the northern dock complex in Liverpool and mixed commercial and residential areas inland from them and, 'across the water' on the Wirral peninsula, Birkenhead's commercial centre, a mixed housing and industrial area surrounding it and leisure and residential areas in New Brighton (see Figure 3.1). The MDC now housed 1,620 businesses employing 31,000 people and 3,215 dwellings accommodating 6,690 people in 2,770 households (Merseyside Development Corporation, 1989).

What is particularly interesting about the extension of the designated area was the influence of 'locality effects' in the form of sustained local political lobbying over the geography of the extended area. The inclusion of a sizeable residential population certainly ran against the grain of central government thinking which was generally reluctant to see Development Corporations extending into residential areas because of the problems that such extensions appeared to have produced elsewhere (interview with Chris Farrow, Chief Executive and Alan Friday, Assistant Director, Economic Development, MDC, December 1997). On the Wirral, the extension of the area around Birkenhead and the rather surprising inclusion of the commercial and residential – but non-dockland – area of New Brighton (a local seaside resort recently fallen on hard times) was lobbied for strongly by the local Conservative MP, Lynda Chalker, who was sitting at the time, as was to be proved in the General Election of 1992, on a particularly slender electoral majority. The inclusion of another large residential population in the Vauxhall area of north Liverpool also bore the clear imprint of 'locality'.

The original designated area of the MDC had a population of only 450 people with, as already noted, 310 of these housed in a single housing estate in north Liverpool. Extension of the area in 1988 increased the population to 7,000. While this figure was relatively small (it was, for example, less than a fifth of LDDC's 40,000 population), it did include some of the area's oldest and most tightly-knit working class neighbourhoods. The people living there had a long history of adversity (in both employment and housing) and a strong sense of place – inner-city Liverpool. This sense of, and commitment to, place had a direct bearing on the MDC's development. Indeed it played a key role in the very redefinition of the Corporation's boundaries.

One of the most active neighbourhood groups was in the Vauxhall area in north Liverpool, historically dominated by dock work and a prominant employer (Tate & Lyle, which closed its factory in 1981). A group of residents, the 'Eldonians', formed a housing cooperative in 1979 to resist dispersal to council housing estates on the city's periphery. Between 1979 and 1983, the cooperative converted rundown tenement flats into accommodation for the elderly and developed plans to build an 'urban village' on the site vacated by Tate & Lyle. These plans were threatened by the election of the 'Militant' City Council implacably opposed to housing cooperatives and committed to a programme of municipal house building. The Council took control of the cooperative's housing stock and refused the planning permission required to change the Tate & Lyle site from industrial to residential use. A bitter and protracted battle between the council and the cooperative ensued and spilled over into intense battles within the local Labour Party (Meegan, 1990). Central government, already at odds with the City Council, certainly made the most of this conflict by financially supporting the Eldonians through the Housing Corporation. But the Eldonians also felt the need to lobby for an extension of the MDC's boundaries into their patch to allow them to escape (with a degree of reluctance given their support for, and indeed in many cases, membership of the Labour Party) from what they saw as malevolent local authority control. When the MDC's boundaries were altered these alterations neatly took in the Eldonian community.

The MDC was now clearly at a turning point, its major reclamation work was virtually complete along with its central 'flagship' project, the Albert Dock. It now had to come to terms with the planning implications of its extension, and particularly with the inclusion of a significant residential population. There were also development pressures challenging its policy preference for mixed commercial and residential uses on the waterfront. These came to a head in early 1989 when the Board opted for a mixed commercial, industrial, leisure and residential development on the Wirral waterfront against a competing proposal for a coal-fired power station and deep river berth. The decision provoked the highly public resignation from the Board of Patrick Minford, then Professor of Economics at Liverpool University, who argued that rejecting the power station scheme meant the loss of an opportunity to secure a new source of manual jobs and thus directly to address the problem of high levels of (male) manual workers. The rest of the Board were less convinced of the estimated job multipliers for the scheme and were worried about the 'blighting' effect that the development would have on Birkenhead and its general environmental impact.

The decision to turn down the power station proposal effectively put the seal on the nature of the waterfront development to be pursued: mixed use favouring commercial and residential schemes in relatively small 'packages' rather than large-scale, single-industry developments. The perceived interest in the relatively small-scale housing developments in the Albert and Wapping Docks also reinforced the push towards residential development that had itself been encouraged by the inclusion of residential areas in the extended boundaries.

Indeed, extension of the area meant that the existing sub-areas needed to be rethought. Five sub-areas were eventually defined – North Liverpool and Bootle; Liverpool Waterfront (including the Albert, Canning, Wapping and Queens Docks); South Liverpool (including Liverpool Riverside and the site of the former Garden Festival); Birkenhead (now including the town's commercial and residential hinterland) and New Brighton (see Figure 3.1). Individual strategies were produced for each of these in 1991.

The extension of the area to include the Eldonian Housing Cooperative in the Vauxhall area of north Liverpool also helped to ensure that the Corporation's housing strategies would also contain a substantial element of social housing. At the time of the extension of the MDC's boundaries, the Eldonians had already won an appeal against the refusal of planning permission for the change of use of the Tate & Lyle site and, with Housing Corporation support, had embarked on the first phase of what was to become the Eldonian Village. The cooperative now had another important sponsor, the Development Corporation with its statutory planning powers. History will clearly show that the Eldonians 'were in the right place at the right time' but, of course, it was their community-based, organisational efforts which got them there.

In 1989, the Corporation set in motion the development of a housing strategy for the whole Vauxhall area developed in consultation with the Eldonians, the Vauxhall Neighbourhood Council (an umbrella organisation for local community groups) and other local groups and businesses. This strategy, the Vauxhall Housing Area Development Strategy, committed the Corporation to a major element of social housing in its residential developments. It also contributed significantly to the resurrection of one of the triad of regeneration programmes identified in its *Initial Development Strategy* – 'social renewal', which had hitherto been overshadowed by the other two programmes, 'physical restoration' and 'economic regeneration' and especially the former.

In terms of 'economic regeneration', there were also the first signs of a strategy towards indigenous business. Business Priority Areas of mixed industrial and commercial uses were designated (two in inner Liverpool and one in Birkenhead) in which small-business coordinators and Business Associations were funded. The focus of this activity, developed alongside programmes for grants and managed workpiece, was firmly on indigenous small firms.

This is not to argue that there was no 'place marketing' for inward investment in this transition period but it had been a rather low-key, 'behind-the-scenes' approach with an emphasis on the public sector and with not unimpressive results. In 1990, the relocations to the MDC area of the Customs and Excise VAT Headquarters and the Land Registry were announced. In the following year, offices for the Child Support Agency were added to the list of future inward investment projects, offering a significant contribution to the achievement of the Corporation's investment and employment targets.

The period also saw an important political development with the return to the Corporation's Board of a representative from the post-Militant Labour-led Liverpool City Council in 1989. This return meant that, for the first time since 1983, all three local authorities were represented, to the evident pleasure of John Ritchie, the Chief Executive of the MDC:

> Merseyside has had this enormous capacity to self-destruct. This is the first time for 15 years when everything has been pointing in the same direction at the same time. (Cited in Hamilton Fazey, 1989)

The Liverpool City Council leader, Keva Coombes, was also clear about the political pragmatism behind the decision to rejoin the Board:

> Politically, we had arguments for some time ... Some councillors were quite strongly opposed on the grounds that it was an appointed, not elected body. But it's there, and certainly will be until the next general election, and maybe even after that. It's had its boundaries extended, giving it planning powers over more of the city, and is getting more influential in housing matters. It isn't going to go away. Far better to be inside arguing our corner. The other argument is that it is better that the development corporation get something, than that the money should stay in the Treasury. It certainly isn't going to come to the city council. That was the clincher. That was the trade union's view as well: better to be on the inside than to pretend it doesn't exist. (*Ibid.*)

The Code of Consultation with the local authorities was revised in consultation with the latter in 1990, promoting the more collaborative approach that was further reinforced politically by the replacement of Keva Coombes by an even more pragmatic Labour leadership in the form of Harry Rimmer in 1990 and Frank Prendergarst in 1996 (who both sat on the Board in its last two years).

The Corporation's expenditure pattern over the transition period reveals the reorientation in its priorities. As Figure 3.2 shows, annual expenditure rose from just over £21 million in 1988/89 to around £27 million in each of the two following years. The continuing dominance of spending on capital works (land reclamation and refurbishment, roads and environmental

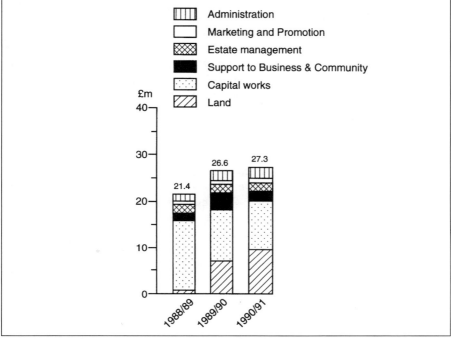

Figure 3.2 *Merseyside Development Corporation: Expenditure, 1988–91*
Source: Merseyside Development Corporation's Annual Reports and Accounts

improvements) in the year the Corporation was extended is clear, still accounting for 72% of total expenditure. This share is much reduced in the two following years, however, with noticeable increases in the share going to land acquisition (26% in 1989/90 and 34% in 1991/2 compared with just 4% in 1988/89) and to support for business and community projects (14% in 1989/90 and 8% in 1990/91 compared with 7% in 1988/89) themselves reflecting the growing importance of land for residential development and training and business support. For the three years as a whole, capital works now accounted for just 70% of the total expenditure compared with its 82% share between 1981 and 1987/88 while support for industry and community (mainly training) saw its proportionate share increase sixfold to 12% from its 2% share between 1981 and 1987/88. In contrast, marketing and promotion still accounted for a relatively minor proportion of spend (averaging around 3% of expenditure over the three financial years).

1992–1998: 'urban entrepreneurialism' Merseyside-style
The appointment, in 1991, of a new Chairman (Desmond Pitcher, the Managing Director of Littlewoods, a major local employer) and Chief Executive (Chris Farrow, recruited from the London Docklands Development

Corporation) set the stage for a change of strategy, not least because the massive investment in land reclamation and building refurbishment of the 1980s had meant that the Corporation had reached a kind of 'regeneration threshold' and that there was a feeling that the 1990s could finally be 'payback time' (interview, December 1997). The strategy involved a proactive 'place marketing' approach and a much more direct engagement with what was seen as a slowly 're-energising' private sector. But this rather belated adoption of the 'entrepreneurial' approach was also necessarily tempered by the recognition that changed political conditions at both national and local level and the social demands of the communities for whom the Corporation was now responsible also demanded a more open, partnership-type, approach. This local 'realpolitik' (and 'realeconomik') was certainly acknowledged by the newly appointed Chief Executive, Chris Farrow, in his reflections on moving from the LDDC to Merseyside: 'You can afford tank warfare in London Docklands when you are two miles away from the Bank of England. You cannot afford it here' (quoted in *The Times*, 13 July 1992, p. 13).

At a national level, the toning down of the pure 'enterprise approach' in urban policy had been revealed with the introduction of City Challenge and the Single Regeneration Budget and the relative downgrading of urban development corporations clearly signalled by the financial cuts announced in the Chancellor's Autumn Statement in 1992 (which, in the case of the MDC, effectively meant a 25% cut in its lifetime budget; interview, December 1997) At local level, there had also been a growing degree of cooperation between local councils, central government and the voluntary sector, not least to benefit from national and European policies, as the three levels increasingly intermeshed. Thus, Liverpool City Council joined with the Mersey Taskforce, Merseyside Tourist and Conference Bureau and the MDC actively to lobby for the relocation of the Customs and Excise VAT Headquarters (interview with Alan Chape, Deputy Chief Executive, Liverpool City Council, December 1997).

To fill the political vacuum created by the abolition of Merseyside County Council, officers in the five local authorities had formed a Steering Group for discussing planning issues at city-region level and this structure was to prove crucial for helping to secure European funding, first from the Merseyside Integrated Development Operation (drawn up in 1988 and subsequently adjusted to 'fit' with the Community Support Framework for North West England in place between 1989 and 1993) and later from the Structural Funds under Objective One (with designation in 1992 and a spending programme from 1994 to 1999). The late 1980s and early 1990s saw a growing partnership between local and central government agencies and the voluntary sector in large part shaped by the catalyst of European funding. And the MDC was to become a key actor in this collaborative activity.

The MDC's gradual turn to a more open, collaborative approach was heralded symbolically, in 1992, by the re-wording of the Corporation's Mission Statement to insert after the verb in the sentence '. . . to secure the self-sustaining regeneration of central Merseyside' the words 'in partnership with others, advances towards . . .'. More practically, for the first time in the history of the Corporation, the meetings of its Planning Board were opened to the public and its Corporate Plans circulated to local authorities and other public bodies.

The MDC thus became more carefully interwoven into an extended network of institutional relations which served, in a sense, to help raise the Corporation's view out of its immediate area and give it a wider Merseyside field of vision. Thus, for example, it became actively involved with Liverpool City Council and Wirral Metropolitan Borough Council in Single Regeneration Budget initiatives in areas adjacent to its designated area (and steered some of its social and community spending in their direction). Its involvement in the Objective 1 programme (including membership of the Monitoring Committee) also meant that its own planning developments could now more clearly be placed in a broader, city-region wide framework.

The Corporation also helped, in 1992, to sponsor the establishment of the 'Mersey Partnership', a public-private body which brings together four of the five local authorities with a range of private firms and the area's three universities to engage in 'image campaigns' and 'place marketing'. It also shifted more money into its own 'place marketing' activities, some in collaboration with the regional inward investment body, INWARD.

In the last financial year before extension of its area, the Corporation had spent some £400,000 on marketing and promotion activities, less than 2% of its total expenditure. And, given its emphasis on tourism and leisure activities, a substantial share of this spending went to support the Merseyside Tourist Board which had lost its main sponsor with the abolition of the Merseyside County Council in 1986. In the early 1990s, the MDC sought to reposition the Tourist Board's activities, encouraging the merger of the Board with the recently created Conference Bureau to create the Merseyside Tourist and Conference Bureau. The emphasis in marketing and promotion, however, shifted perceptibly towards attracting private sector investment with North America and the Far East providing the focus. This shift was inaugurated with a high profile trade mission to the United States with the Royal Liverpool Philharmonic Orchestra in tow. It was strongly felt that 'place marketing' of the MDC area, and Merseyside more generally, in the US and Far East would encounter less negative 'image' problems. In terms of attracting private sector investment, as the Chief Executive, Chris Farrow, puts it: 'It was easier to get Costco [a US discount warehouse] than Sainsburys' (interview, December 1998).[1]

The inward investment strategy was also strongly influenced by the geography of the designated area which lacked sites suitable for projects requiring a large land area. The 'Nissan North East', 'Toyota Derby' type projects could not be accommodated in the designated area. By the end of 1996 the promotional activity had managed to attract four major inward investment projects from the US and one from Taiwan. With another two from the UK and one from Canada, the projects together brought with them an estimated 770 jobs and some £21.6 million in investment. The scale of this investment is perhaps less significant, however, than the way in which attempts were made to attach it to targeted local employment. The location of Costco, the north American-owned discount warehouse shopping facility on the former Tate & Lyle site in Vauxhall, north Liverpool, provides the best example. The MDC, in partnership with the Employment Service, Merseyside TEC, Liverpool City Challenge and a number of local voluntary groups, set up the so-called 'Costco Employment Initiative'. The aim was to maximise the recruitment of inner-city residents through a coordinated programme of information, counselling and training (including training in interview skills).

Local community groups produced their own promotional material and put up posters, leafleted local homes and community facilities and organised a series of 'roadshows' using a specially produced video of the job opportunities. The end result served to underline the difficulties of such targeted employment exercises. The company was swamped with applications, the 161 jobs in the initial recruitment phase attracting 2,850 applications – itself a clear reminder if one were needed of the depressed state of the local labour market. The Employment Initiative itself worked initially with 248 local residents of which 88 went on to pre-interview training. Seventy-eight of these were offered interviews at which 70 turned up. Eighteen of these were offered jobs and 13 held on a reserve list. The Initiative thus secured just 11% of the initial recruitment (although, in total 27% of Costco's local workforce was previously unemployed). But important lessons were learned, including the need to give community groups in such partnerships more active roles for formulating and managing them as well as being the main routes through which to target potential employees (for an evaluation of the Initiative, see Russell, 1996).

Another inward investment project, this time from elsewhere in the UK and by the public sector, also showed how equal opportunities policies can be integrated into such local employment initiatives. The inward investment in question was the relocation of the Customs and Excise VAT Headquarters from Southend to the Queens Dock, south of the Albert Dock on the Liverpool Waterfront. The transfer had been announced, with great fanfare, by the then Secretary of State for the Environment, Chris Patten, in October 1990. With the promise of over 1,000 jobs it was seen as a great coup by the MDC and the central and local government partnership that had formed to lobby for it. In the event, internal reorganisation, the fall of VAT registration business and the need to accommodate more than expected from the existing Customs and Excise operation in Bootle, north Liverpool, significantly limited local recruitment potential.

Nevertheless, local recruitment was necessary and crucially the local targeting of this recruitment was reinforced by the fact that the relocation, in 1994, coincided with a drive to increase the representation of ethnic minorities in the civil service (as spelled out, for example, in the White Paper,

The Civil Service, Continuity and Change, Cmnd 2627, published that year). To meet the requirements of equal opportunities the recruitment programme could not avoid having to engage with the black population of Liverpool 8. Ironically, the Dean of Liverpool Cathedral, himself heavily involved in local economic regeneration activities, had attempted unsuccessfully to have this population brought within the extended boundaries of the MDC 1988. Now, a major civil service employer, located less than a mile away from the heart of Liverpool 8, was committed to a recruitment campaign directly targeting local ethnic minorities. Its attempts to implement it ran straight into the formidable barriers surrounding the local recruitment of black Liverpudlians, barriers erected by a long history of racism in local economic, political and social life. The difficulties were revealed not least in the outcome with only six black people being recruited after four years of careful effort (for an authoritative history of the campaign, see Moore, 1997). But, despite the initial small gains, the recruitment exercise, like the Costco Employment Initiative, not only proved an important learning experience for both the employer and local community groups it was also critically symbolic. Local employment initiatives – with positive action on equal opportunities – were now clearly on the local regeneration agenda.

The Costco Employment Initiative also coincided with a shift in the MDC's training activities towards more small-scale, community-based projects. The Corporation had not been involved in any significant way in training until 1986 when it stepped in to help secure funding for a major training provider, Merseyside Education Training Enterprise Limited (METEL) with the abolition of the latter's main sponsor, the Merseyside County Council. The company was reconstituted as a public-private partnership with charitable status and with the MDC holding a controlling interest. METEL received about 53% of the MDC's training expenditure between 1986 and 1992 (a total of just over £2 million) but, despite this support (and that of the Manpower Services Commission and the European Social Funds that the MDC's funding attracted as 'match') it ceased trading at the end of 1992. This collapse, which coincided with the extensive reorganisation of training provision that accompanied the establishment of Training and Enterprise Councils, certainly revealed the dangers of the Corporation having most of its training eggs in one basket and encouraged a shift in emphasis towards supporting a range of relatively small, customised training initiatives with a local community focus and encouraged, at the same time, by the broader level shift towards social and community activities that extension of the area had engendered. In its Corporate Plan in 1992 the Corporation recognised the need for a 'Community Programme':

> To complement the economic regeneration of the Designated Area with a comprehensive programme of community support. This programme will be developed in consultation with local community organisations and local authorities . . . the emphasis of the programme will be addressing the problems of young people with the provision of new facilities, improvements to existing facilities and support for

innovative approaches to youth work . . . support will be directed at assisting the development of community based economic development initiatives which can become self-sustaining and which will complement the redevelopment of the area for social housing predominantly. (Merseyside Development Corporation, 1992a, para. 5.23)

A natural focus for the social programme was the Eldonians' social housing scheme and, in addition to land preparation, grants were given for a nursery and community centre. Support for a nursery at the nearby Vauxhall Neighbourhood Council and for the establishment of another housing cooperative, the Vauxhall Housing Cooperative. By the mid-1990s community support shifted away from physical infrastructure towards education and, as already noted, training. Again, much of this was concentrated on the two largest community organisations in the area (the Eldonians and Vauxhall Neighbourhood Council in north Liverpool) and included, for example, grants for care training of both children and the elderly. Grants were also given to support general training in community-based economic development as well as specific initiatives (like the Furniture Resource Centre in South Liverpool which builds community-focused training around the recycling and sale of secondhand furniture).

Harvey (1989a and 1989b) argues that one defining feature of 'urban entrepreneurialism' is the growing use of the 'urban spectacle' including the hosting of one-off events (to attract the 'consumer dollar' in Harvey's terms) and it is interesting that it was precisely such an event that provoked the Corporation's most public censure. The event in question was arranged in 1992 around the return visit to Liverpool of the fleet of sailing ships, the so-called 'Tall Ships', that had called in on the city in 1984. This time the visit was part of the global commemoration of the 500th centenary of Christopher Columbus' North American foray. As part of the welcome to the ships, a high-profile concert of classical music and opera was held on reclaimed dockland with the King and Queen of Spain as special guests. The MDC handed over the organisation of the concert to an off-the-shelf promotions company. But in a city in which the Royal Philharmonic Orchestra and the 'arts' in general struggle for public sponsorship, the event was, to say the least, always likely to be risky in terms of public support at the box office.

And so it proved, with the MDC having to step in at the last minute to take up and distribute unsold tickets and meet other expenses to allow the 'spectacle' to take place. The promotions company went bankrupt, leaving behind a trail of creditors.[2] And, in due course, the Corporation found its involvement in the financial débâcle coming under the scrutiny of both the National Audit Office and the House of Commons Select Committee. The National Audit Office Report (National Audit Office, 1994) was highly critical, finding the Corporation guilty of 'unauthorised spending'. The MDC was duly 'fined' some £295,443 – a figure which was deducted from its grant-in-aid for 1994/5 in a reminder, if one was needed, of where the political accountability of Development Corporations firmly lay.[3] Chris Farrow,

Chief Executive of the MDC, certainly felt that the whole episode did raise some interesting issues of public accountability (interview). In this view, the fact that Development Corporations were unelected bodies inevitably meant that their promotional events would be controversial.

In support of this argument, he pointed to the fact that the 'Tall Ships' events, organised by the local authorities in both Newcastle and Portsmouth, received no parliamentary scrutiny, despite the fact that they cost 'twice as much' as the one on Merseyside. It was also significant, he felt, that the criticism of the Merseyside event came from central government but not from the Merseyside local authorities, who were supportive – a contrast which raises an interesting questions about appropriate levels of public accountability for localised bodies such as the Development Corporations. Chris Farrow also felt that the parliamentary inquisition was also part of a broader political attack on the remuneration of directors and chief executives of newly privatised public utilities. In Chris Farrow's opinion, the MDC provided a prime target for such an attack as its Chairman, Desmond Pitcher, had been anointed by the media, and unfairly so, as 'king of the fat cats'.

In the same year's Financial Statements (Merseyside Development Corporation, 1994), reference was made to the determination by the Secretary of State for the Environment of the Corporation's Termination Date of March 1998. Somewhat ironically, in some senses, this date was announced just as the 'payback' that Chris Farrow had alluded to was starting to happen with public-private investment 'leverage' ratios beginning to reach 1:4 and the sale of land bringing sufficient income to compensate for reductions in grant-in-aid. This income certainly represented a major change in the local property market. At the beginning of the 'entrepreneurial' period (June 1992) the MDC still held some 87% of the land that it had acquired (333 out of 382 hectares) and had a negative balance on land transactions (of nearly £18 million with expenditure of £29 million and receipts of £11.2 million). This position was in marked contrast with that of its contemporary, the LDDC, which held only 68% of its land acquisitions (594 out of 870 hectares) and had been able to make a surplus on its land transactions (with a spend of £161 million and receipts of £302 million giving a positive balance of £141 million) (Hansard, 1992a).

The impact of the Corporation's housing strategies was also starting to become visible with the retention and extension of social housing (through housing association developments) in North Liverpool and Birkenhead, the creation of relatively low-cost private sector housing in existing residential areas of Birkenhead and New Brighton and the establishment of entirely new private housing developments on the Liverpool Waterfront and in South Liverpool.

In terms of social housing, the biggest single impact was in North Liverpool with the second phase of development of the Eldonian Village (completed in 1995 and now comprising 295 dwellings and a residential care home for the elderly) and the establishment on land reclaimed by the MDC of the nearby Athol Village Housing Cooperative (with 150 new homes). The main housing impact in Birkenhead was the development, again through local housing associations, of relatively low-cost housing and particularly in the form of apartments over shops in specially designated Shopping Improvement Areas (the so-called 'living over the shop' scheme). A similar scheme was also introduced in New Brighton along with a mix of private 'luxury' and 'non-luxury' housing.

The most important private sector developments, however, were on the Liverpool Waterfront (with the continuing development of luxury apartments in the Albert, Wapping and Waterloo docks and flats and apartments – initially for rent but more recently for owner occupation – in the docks surrounding the nearby Marina) and in the former Garden Festival area of Riverside (with the development of 'luxury' family accommodation). Together, these housing developments have certainly had a significant impact on residential geography and housing tenure in the MDC area, as Tables 3.3 and 3.4, using figures from the assessment of housing outputs commissioned by the MDC (ERM Economics, 1997a), clearly show. The tables give figures for the MDC area at its extension in 1988 and eight years later. The area has seen household and population numbers more than double over the eight year period, as Table 3.3 shows. But, as the table also shows, this overall increase disguises substantial geographical variation including, most noticeably, the creation of virtually new private housing areas on both the Liverpool Waterfront and in South Liverpool.

The predominantly housing association-sponsored developments in North Liverpool and Bootle have also produced a less dramatic but still relatively significant increase in both housing and population in those areas. The changes have also had a significant impact on housing tenure in the area, as Table 3.4 shows. In line with national trends is the shifting balance in the provision of social housing from local authorities to housing associations, with the former seeing a halving and the latter a quadrupling of their respective shares of the housing stock between 1988 and 1996. The relatively small increase in the private sector share (from 60% to just under 64%) does put into perspective the relative scale of private housing developments in the wider MDC area. But new housing areas have been created in a city-region currently losing population at a faster rate than any other in the United Kingdom, a piece of 'residential social engineering' that I will return to below.

The period also saw the inauguration of two major 'flagship' projects: the Twelve Quays International Technology Campus on the Birkenhead waterfront and Atlantic Avenue in North Liverpool. 'Twelve Quays' involves the development of 11 hectares of derelict land for pharmaceutical and related high-technology companies and 'spin-off' research activities from Liverpool John Moores University, including, in collaboration with the Royal Greenwich Observatory, what will be the world's largest robotic telescope. The rest of the site will house a terminal for a 'roll-on, roll-off' ferry link with Ireland. 'Atlantic Avenue' is a major development (costing in total some

Table 3.3 Percentage distribution of housing stock by strategic sub-area, 1988 and 1996

	Private		Housing Association		Council	
Sub-area	1988	1996	1988	1996	1988	1996
North Liverpool & Bootle	3.8	16.1	1.8	50.8	94.4	33.1
Liverpool Waterfront	12.5	89.5	0.0	6.4	87.5	4.1
South Liverpool	75.0	85.5	1.6	13.0	23.4	1.5
New Brighton	92.6	90.6	4.4	7.0	3.0	2.4
Birkenhead	45.9	49.4	8.4	19.7	45.7	30.9
Total MDC Area	60.1	63.7	5.1	20.8	34.8	15.5

Source: Based on figures in ERM Economics (1997a)

Table 3.4 Change in households and population by strategic sub-area, 1988–1996 (Index 1988 = 100)

	Households			Population		
Sub-area	1988	1996	Index 1988 = 100	1988	1996	Index 1988 = 100
North Liverpool & Bootle	398	1,129	284	1,007	2,540	252
Liverpool Waterfront	26	518	1,992	50	929	1,804
South Liverpool	68	1,009	1,484	201	2,601	1,294
New Brighton	1,012	1,257	124	2,064	2,497	121
Birkenhead	724	1,068	148	1,549	2,126	137
Total MDC Area	2,228	4,981	224	4,871	10,692	219

Source: Based on figures in ERM Economics (1997a)

£40 million) which aims to unify the North Liverpool and Bootle strategic sub-area by the upgrading of the main route through it from the city centre. The development involves a mix of environmental works, the preparation of sites for industrial and commercial use and improvement grants for existing businesses. Together the two projects represented the Corporation's last major development schemes.

The last years of development work was undertaken in the context of a significant repositioning of the Corporation in the local policy arena as a result of the approval of the Objective One 'plan' (more accurately 'single programming document') for Merseyside. This £1.2 billion spending programme for the period 1994–1999 not only meant that, politically, the MDC was no longer 'the only regeneration show in town' but was also able to position its development activities in a broader strategic planning framework for Merseyside as a whole – a framework that has certainly been notable by its absence in its London contemporary. For the MDC: '50% of the benefit of Objective One is its plan and strategy . . . and its focus on priorities . . . [in which] . . . everyone knows where they stand' (interview, December 1997). With representation on the Monitoring Committee, and advisory sub-committees, the MDC was able to relate its own priorities to those set out in the Objective One programme and its developments could be placed in a developing 'plan' for the city-region as a whole. Consequently, in its final years, the MDC became just one, albeit important, piece in a broader planning jigsaw for Merseyside. Indeed, the MDC itself can take some credit for the part it has played in the resurrection of planning on Merseyside.

As the Corporation notes in its valedictory *Regeneration Statement* (Merseyside Development Corporation, 1997c), it was established at a time when there was no comprehensive or up-to-date statutory planning framework for its designated area. Five years into its life, the Merseyside County Council was abolished, provoking the inevitably slow formation of the 'shadow' but non-statutory planning framework that emerged from the voluntary collaboration of the city-region's five constituent local authorities. The Development Corporation's *Initial Development Strategy* (MDC, 1981) was the only detailed, up-to-date land-use plan for much of its designated area and major elements of this strategy were to find their way into the Unitary Development Plans of the three local authorities in its area. As already noted, it helped to support the establishment of Single Regeneration Budget Partnerships operating in parts of its area and it played an important role in the development of the Objective One spending programme. Planning has reappeared on the Merseyside agenda and the MDC can be credited with an important contribution to this.[4]

Its withdrawal from the planning scene does also appear to have been relatively orderly. Discussions about succession with successor bodies (the three local authorities, the Government Office Merseyside, the Commission for New Towns, English Partnerships and various management companies) were begun early in 1997. Consultants were commissioned to produce appraisals of the Corporation's achievements in relation to both key outputs and major projects in the first half of 1997.[5] The reports inform the museum exhibition and accompanying brochure (*All Our Tomorrows*), already referred to. The Corporation also produced, in consultation with the various successor bodies, its *Completion, Exit and Succession Plan 1997/98* and detailed *Regeneration Statement*, with the latter setting out in 'workshop manual' form the regeneration opportunities remaining after its demise in 1998. A clear aim, and one subject to considerable local media attention, has been to keep the remaining monies on Merseyside and not allow them simply to be siphoned back into the Treasury's coffers. Thus, a package of targeted projects has been brought together, in negotiation with successor bodies, to ensure that sufficient funding 'dowries' remain to enable the projects to be completed and maintained. The sale of the Corporation's landholdings to the private sector was also accelerated, leaving one half of the 496 acres of land to be transferred to successor bodies already subject to contracts or with terms agreed with private sector developers.

Figure 3.3 shows the Corporation's changing patterns of expenditure between the financial years 1991/92 and 1996/97. Total expenditure peaked in 1992/93 (at £39.2 million), the year when the major cuts in grant-in-aid were announced. Although expenditure did fall thereafter, it is noticeable that it was still maintained in the last four years at an average (£28.3 million) above that for the previous period of transition between 1988/89 and 1990/91 (which saw average expenditure of £25.1 million). Internally generated receipts from the sale of land and other assets – the 'payback' referred to by the Chief Executive, Chris Farrow (and an indication of a slowly reviving local economy) – were able to sustain these levels of expenditure, rising from nearly a quarter of the total expenditure in 1993/4 to just over 39% three years later (MDC, 1997d). Land acquisition retained its relative share of expenditure in the first three years but fell off in importance thereafter as development programmes were completed. The growing community focus of expenditure can also be seen, especially after 1991/92 when it reached an average annual figure of just under £5 million and actually grew in relative importance as total expenditure declined (especially in 1994/95 when, at nearly £6 million, it accounted for 25% of total expenditure). What is also noticeable is the absolute and relative decline of expenditure on marketing and promotion after the peak expenditure in 1992/93 (just under £4 million, approximately 4% of total expenditure), the year of the Grand Regatta financial débâcle. This decline reflects in part the nature of such expenditure (what

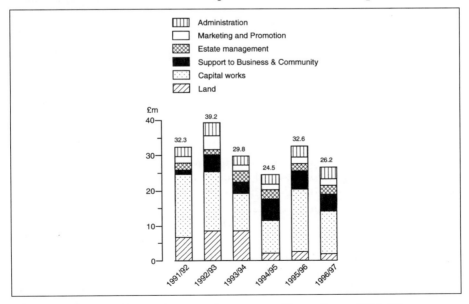

Figure 3.3 *Merseyside Development Corporation: Expenditure, 1991–97*
Source: Merseyside Development Corporation's Annual Reports and Accounts

Chris Farrow described as 'promotional froth'). With lead-times of between three to five years between promotional activity and actual projects on the ground, it was clear that such levels of expenditure could not be sustained by a body facing closure. Emphasis, in terms of promotional activity, thus switched after the peak year of 1992/3 'wholeheartedly to partnership in inward investment with Mersey Partnership and Inward' (interview).

The MDC in retrospect

The three periods described above thus saw the operations of the MDC shaped by the local articulation of a set of economic, political and social influences at local, national and international scales. Dominating the economic influences was the depressed state of the local economy. The MDC's seventeen year lifespan traversed a national economy experiencing two complete economic cycles of recession and growth. It was only in the last few years of the MDC's life that any signs of local economic revival, however modest, were visible. Local development possibilities, especially in industrial and commercial property, were restricted accordingly.

Politically, the Corporation also operated against a changing backcloth of national urban policy. Introduced initially as one of the two Development Corporation 'flagships' of the 'enterprise approach' of the Thatcher governments of the 1980s, this privileged position was eventually undermined by the waning of enthusiasm for the approach in the 1990s. This policy change was precipitated in part by the growing exchequer costs of the Corporations (greatly influenced by the costs associated with the London Docklands Development Corporation – which provided a clear 'London-locality' effect on national policy) and also by a tipping of the balance in urban policy between 'property-led' and 'people-based' emphases towards the latter.

The interaction between national and local politics also played a crucial role in the MDC's development. The confrontation between Liverpool City Council and the Thatcher government of the mid-1980s inevitably left the MDC, perceived as an arm of central government policy, politically isolated. The Labour leadership of the post-Militant council instituted a reconciliation, at first pragmatic but later more constructive. And this developing cooperation was further encouraged by both the shift in national urban policy towards 'partnership' and the local social and community issues that extension of the MDC's area in mid-life had placed on the agenda. Cementing this shift was the Corporation's gradual assimilation into the broader planning framework and institutional networks that European policy, in the form of the Merseyside Integrated Development Operation and most notably the Objective One spending programme, encouraged.

In the first years of the MDC's life, urban entrepreneurialism was not on the political agenda. A fragmented local political base and a weak private sector undermined the development of the political coalitions that urban entrepreneurialism requires. The MDC thus operated in its first years as a relatively isolated environmental and land reclamation agency with only limited attempts at 'place marketing' and these mainly aimed at attracting visitors to the slowly restored waterfront. Gradually, as the land reclamation programme neared completion, the MDC switched to a more 'entrepreneurial' approach in relation to property development and attracting inward investment both on its own initiative and as part of the slowly developing public-private sector partnership that it was itself active in sponsoring. Reinforcing this partnership approach was the catalyst of European regional development funding programmes. In the Corporation's final years, then, its property-development and place-marketing entrepreneurialism was clearly tempered by the 'partnership' approach and social and community focus engendered by local, national and international political and policy developments.

The MDC: some general lessons

This is not the place for a detailed project by project evaluation of the impact of the Corporation's activities on the local economy.[6] What I would like to do, instead, is to draw out from the MDC's activities, a number of broader lessons for urban regeneration policy more generally.

The overriding message of the MDC's activities on Merseyside has to be the continuing need for a massive and sustained public-sector-led intervention in the regeneration of such disadvantaged city-regions. Between 1981 and the end of the its final financial year of operation, the MDC had spent some £500 million, of which £399 million (nearly 80%) was funded by central government grant-in-aid. The remaining balance of £101 million was financed partly by the European Regional Development Fund (about 10%) but mainly internally generated receipts, the latter being, as already noted, only really significant in the last four to five years. Certainly, the ratio of internally generated receipts was on a marked rising trend with central government grant only accounting for some £20 million, or 30% of the final year's £69 million expenditure. In terms of two of the key 'standard outputs' set for Development Corporations, this spending has generated some £668 million in private sector investment and just over 20,000 (gross) jobs.[7] A crude comparison of these outputs with those of other UDCs would certainly, at first sight, appear relatively unfavourable particularly in terms of public:private investment 'leverage'. Pike (1997) provides figures for such a comparison.

On the basis of 'lifetime' figures for twelve UDCs, the ratio of grant-in-aid to private sector investment ranges from 1.0:0.43 (Plymouth DC) to 1.0:7.9 (Birmingham Heartlands) with an overall total of 1.0:3.8. The MDC is ranked eleventh, with a ratio of just 1.0:1.8; the private sector investment of its later years clearly being unable to outweigh the massive public investment in reclamation and environmental refurbishment of its first decade of operation. In terms of 'lifetime' grant-in-aid cost per (gross) job, figures range from £5,733 (Sheffield DC) to £90,631 (Plymouth DC) with an average for all twelve UDCs of £16,873. The MDC is ranked sixth with a grant cost per job just below the average at £16,496.[8] Such simple comparisons are deceptive, however, because they fail to take into account the differing social and economic circumstances in the localities in which the UDCs are operating. The MDC was introduced in a city-region which was in profound decline on all economic and social indicators. The first ten years of its operation saw gross domestic product per head plummet towards the level that triggers intervention under Objective One of the EU's Structural Funds (75% of the EU average). Between 1989 and 1991, the year before such designation, Merseyside lost some 68,000 jobs, a decline of over 12% (when jobs grew nationally by over 4%). The years since then have seen decline continue, albeit at a slower rate.

The enduring weakness of the local economy – and the concomitant need for regeneration efforts to be sustained – are indicated in Table 3.5, which compares unemployment rates in Government Office Regions over the last five years. It needs to be emphasised that the period is one in which the bulk of the MDC's investment and employment 'outputs' have been achieved and one in which the Objective One spending programme has been operative (1994 onwards). The period was also one of substantial national growth which saw official unemployment rates fall from 10.3% to 5.5%. Unemployment rates also fell in all the Government Office Regions but it is significant that this fall was by far the smallest on Merseyside which, as a consequence, saw its unemployment 'relativity' already the highest in 1993 (at 148, with the UK equal to 100) increase steadily (to 193 in 1997) to remain the highest amongst the thirteen regions.

These depressing employment and unemployment figures certainly put the MDC's job creation figures into sharp perspective. As already noted, the Corporation claimed to have created just over 19,000 jobs in its Designated Area by the end of 1997. It should be emphasised that the figure relates, as required by the Department of the Environment 'UDC Guidelines', to 'gross' job creation. The figure also does not measure any 'job shrinkage' from the initial employment targets reported by employers.[9]

The Corporation, to its credit, did commission annual surveys of employment in its area after 1990 (the absence of such surveys prior to this date itself revealing the relatively low priority given to employment creation). What these surveys show is that the 'gross' job creation over the period managed to maintain employment levels in the extended Designated Area at around 32,000 jobs (ERM Economics, 1997b). So employment in the Designated Area, with all the latter's privileged spending, has been kept stable against the backcloth of continuing employment decline at the level of the city-region. The shift in the locus of employment back to city and waterfront area is clear. So too is the cost of employment generation in such a depressed locality. Jobs were maintained in a Designated Area which accounts for just 1.5%, 0.5% and 7.4% of, respectively, the land area, population

Table 3.5 Unemployment rates and relativities for Government Office Regions, 1993–1997

Government Office Regions	Unemployment rates (claimant unemployed as % of workforce)				
	1993	1994	1995	1996	1997
North East	13.0	12.4	11.5	10.6	8.2
North West	9.5	8.7	7.6	6.9	5.0
MERSEYSIDE	15.2	14.9	13.7	13.1	10.6
Yorks & Humberside	10.4	9.7	8.7	8.0	6.1
East Midlands	9.6	8.8	7.7	6.9	4.9
West Midlands	10.9	9.9	8.4	7.4	5.5
Eastern	9.4	8.1	6.9	6.1	4.2
London	11.6	10.7	9.8	8.9	6.7
South East	8.6	7.3	6.2	5.4	3.5
South West	9.5	8.2	7.1	6.3	4.2
Wales	10.4	9.4	8.8	8.2	6.1
Scotland	9.9	9.4	8.2	8.0	6.4
Northern Ireland	14.1	12.7	11.4	10.9	7.9
UK	10.3	9.4	8.3	7.6	5.5
	Relativities (UK = 100)				
North East	126	132	139	139	149
North West	92	93	92	91	91
MERSEYSIDE	148	159	165	172	193
Yorks & Humberside	101	103	105	105	111
East Midlands	93	94	93	91	89
West Midlands	106	105	101	97	100
Eastern	91	86	83	80	76
London	113	114	118	117	122
South East	83	78	75	71	64
South West	92	87	86	83	76
Wales	101	100	106	108	111
Scotland	96	100	99	105	116
Northern Ireland	137	135	137	143	144
UK	100	100	100	100	100

Note: The unemployment rates for 1993 to 1996 are annual averages. The 1997 rate is for June 1997.
Source: Labour Market Trends, December 1997

and employment of Merseyside County and one which was privileged to receive some £431 million in expenditure from its Development Corporation. What this geography of employment creation also appears to do, is put another nail in the coffin of 'trickle down' in urban regeneration. In Myrdalian terminology, the 'backwash effects' of growth in the MDC area appear to be clearly outweighing any 'spread effects' – unless, of course, we are prepared to wait until that Keynesian long-term in which we are all dead.

There is a need, therefore, for regeneration strategies within urban areas to be linked, as the MDC itself came to recognise, to broader planning frameworks and strategies for the social and economic regeneration of the wider metropolitan regions in which they are located. On Merseyside, the Objective One single programming document is a step in this direction. The proposed Regional Development Agency for the North West might also take things a step further by linking sub-regional plans to regional ones (although the 'toothless', consultative role currently envisaged for the Regional Chamber suggests that this linkage may be undermined, in terms of political account-ability, by a regional-level 'democratic deficit').

Another lesson from the clear weakness of 'trickle down' is the need for regeneration efforts to be sustained and not, as appears to have been the case to date, to take the form of a set of ad hoc, time-limited policy exper-iments. Alan Chape, the Assistant Chief Executive of Liverpool City Council, certainly feels that the MDC was itself a victim of such a 'short-termist' per-spective:

> you can argue about whether or not there should be a special agency in the first place but once you've set one up then the idea that you finish it before it's fin-ished the job is an interesting policy issue. The whole *raison d'être* of the thing was supposed to be its ability to have this single focus . . . so you set it up and it has seventeen years. It still probably needs another five and you kill it off – but it hasn't finished the job. So you then have to create a whole new set of arrange-ments, the Councils, English Partnerships, the Commission for New Towns etc. There's an issue how we conceive regeneration programmes as these time limited things when maybe we should conceive the programme as 'well there's a job to be done and we'll finish when we've done the job. (Interview, December 1997)

The key elements of any comprehensive strategy for social and economic regeneration are captured in Darwin's (1990) 'ESCHER' acronym for an holistic regeneration approach which integrates economic, social, cultural, health and ecological initiatives. A strictly market approach will only address narrowly-defined economic issues. As already argued, the MDC certainly found its property-development approach overshadowed by the costs of tack-ling the environmental legacy of previous development in its area and its policies inexorably tugged in the direction of social evaluations (in, for exam-ple, local pressures for social housing and community-based training and employment policies).

In terms of the environment, its experience certainly raised important issues about the costs and nature of land reclamation. Its Designated Area was, of course, one in which, as already emphasised, environmental degra-dation was at its worst.[10] And, as Figure 3.4 shows, the bulk of the Corporation's expenditure was duly accounted for by physical works and particularly land reclamation and refurbishment (£178 million, 41% of total spending), roads (£55 million, 13% of the total) and environmental improve-ments (£33 million, 11% of the total). Together these physical works accounted for nearly 62% of total expenditure and clearly overshadow the other categories. In comparison, grants to business and community projects account for some £34 million of the total expenditure (just under 8%), revealingly more than double the spend on marketing and promotion (£16 million, less than 4% of the total) but still slightly less than the expenditure on administration (£38 million, nearly 13% of the total).

The physical works related to the degraded local environment – the so-called 'abnormal costs' that need to be removed before developers will develop. But where do these 'abnormal costs' originate? In the case of the MDC, from the private sector and the public sector using private market calculations.

Take, for example, the MDC's reclamation of the 250 acre site for the International Garden Festival. Thirteen million pounds were spent on what at the time was the largest single reclamation project in Europe. What was being reclaimed? The private sector donated derelict oil tank farms (previously operated by Esso and Texaco) and a graveyard of disused petrol-storage tanks (courtesy of BNOC). The public sector chipped in an unstable, methane-generating refuse tip, while the public-private restructuring of the South Docks bequeathed the silted-up Herculaeneum Dock. (The future costs of the reclamation of the Albert Dock complex were also, of course, crucially affected by the decision in 1972 not to meet the expense of securing and maintaining the South Dock system.)

What this all adds up to, of course, is a local manifestation of the broader failure of economic policy to address seriously, to use the title of Kapp's sadly neglected classic, the 'social costs of business enterprise' (Kapp, 1978) or the complex issue of 'sustainable development' (World Commission on Environment and Development, 1987; Redclift, 1987). To insert 'local' into the latter clearly requires the support of both national and international public-policy intervention – and it will be an expensive exercise.

Tackling these environmental 'social costs' also raises issues about development uses of the reclaimed land. In the MDC, a significant element in the

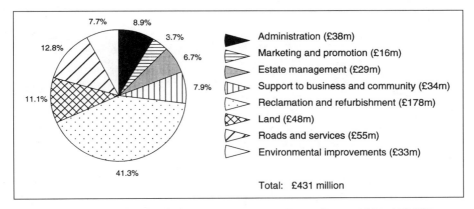

Figure 3.4 *Merseyside Development Corporation: Expenditure, 1981–1996/97*
Source: Merseyside Development Corporation's Annual Reports and Accounts 1996/97

development activity was in the area of commercial leisure uses, with tourism being a key focus. Indeed, the MDC's emphasis on the tourist potential of the Albert Dock can be traced back to the late Merseyside County Council's 1980 Structure Plan and earlier proposals for the development of the waterfront for tourism. And it is difficult to disagree with the MDC's argument that the Garden Festival and the subsequent development of the Albert Dock have provided the building blocks of a tourist industry on Merseyside (see figures 3.5 and 3.6). The MDC claims, for example, that some 3.4 million people visited the International Garden Festival generating an increase of £12 million in local income. The Albert Dock, it is also claimed, has attracted on average 5 million visitors annually since its official opening in 1988 and

Figure 3.5 *Albert Dock before*

Figure 3.6 *Albert Dock after*

£482 million in total visitor spending. The controversial Grand Regatta Columbus in 1992 was estimated by the MDC to have been visited by 2.5 million people who spent some £17.6 million locally.

As with all estimates of tourist 'outputs', some caution needs to be retained over the accuracy of these figures (especially the visitor figures for the Albert Dock.)[11] It is unquestionable, however, that these events and attractions have 'opened' Merseyside, and especially Liverpool and its waterfront, to a substantial number of visitors from outside Merseyside who have spent money locally. The tourist industry that these developments helped to create has in turn also generated a number of jobs locally – relatively low paid perhaps, but jobs nonetheless. Perhaps more important, however, is the way in which the developments have opened up the waterfront to local people.

Here was a city with limited public access to its waterfront (largely restricted, on the Liverpool side of the Mersey, to the ferry terminals, bus terminals and limited recreational facilities in front of the 'Big Three' waterfront buildings). Here also, in the form of the Albert Dock complex, was the largest single grouping of listed Grade 1 Buildings of Architectural and Historical Interest in Great Britain, in a state of total dereliction. The Garden Festival, with its waterfront public walkway and viewing points, the Albert Dock with its leisure and cultural facilities and the restoration and redesign of the Pier Head transformed this situation. From the groups of people who regularly gather on the riverside at weekends and on public holidays to picnic and watch the river flow to the young children who, much to the evident displeasure of some of the shop owners, appropriate the Albert Docks as public swimming pools in the summer months, the river has become a social focus, a place for local residents to visit and, importantly to take visitors. The 1996 visitor survey, cited in KPMG (1997a), showed that 50% of the visitors to the Albert Dock were from Merseyside and local newspaper surveys have confirmed the popularity of the Albert Dock as a leisure and recreational facility for local residents and giving some support to the view of the Chief Executive of the MDC, Chris Farrow, that Albert Dock is much more of a popular icon than the Canary Wharf complex in London Docklands (interview, December 1997).

Underlying all these considerations are questions of the definition of 'tourists' and 'visitors' and the recognition that the object of the 'tourist's gaze' (Urry, 1990) also comes under the purview of the 'resident's gaze'. The MDC, in its environmental programme has unquestionably secured some important recreational, cultural and educational assets, like the museums and Tate Galley in the Albert Dock, that are assets for the city-region and its residents as much as they are for 'tourists' and need to be judged – and funded – accordingly.[12]

In this context, it is illuminating to set the financial débâcle of the grand 'image spectacle' of the Grand Regatta Columbus against the much more popular – and cheaper – Summer Pops of the Royal Liverpool Philharmonic Orchestra, hosted annually for the last few years in the Kings Dock, or the fireworks displays on the river. And the MDC's more low-key sponsorship of local festivals (like the Brouhaha street theatre and the Festival of Comedy) which, by chiming in with burgeoning local arts and culture industries, seem to offer more for the residents of the city-region than the grand 'tourist spectacle'.

A significant proportion of the reclaimed land was also used for housing and, as argued earlier, has resulted in a considerable 'social engineering' of housing tenure and pattern in the designated area. Much is often made of the social-spatial polarisation that Development Corporations have produced in their housing developments through untrammelled gentrification, particularly in London Docklands. The picture on Merseyside is not quite so straightforward. The housing developments here have involved a range of housing tenures – from cooperative and social housing to private housing for rent and sale (across the price range from 'first-time' rented or bought properties to luxury apartments and family houses).

The new private housing on offer has clearly provided, on the basis of sales, an attractive option to middle-class residents who had previously been leaving the city and city-region in droves – with all the implications that this departure has for the local income and tax base. It is hard to exaggerate the importance of retaining these groups in a city which, according to the latest Index of Social Conditions, is still the most disadvantaged in England and Wales. The new private housing has also been introduced in previously non-residential areas and, in most cases, on previously derelict and disused land – avoiding, as a consequence, the immediate disruption and displacement of existing residents. The visible success of these sites, in terms of sales, also gives some credence to the policy of the current government to encourage housing on 'brownfield' inner-city sites. Sites do not come much more 'brownfield' than those developed by the MDC. And not all of these sites are for middle-class residents.

The cooperative housing in North Liverpool in the form of the Eldonian and Athol 'urban villages' also shows that cities can retain working-class residential communities in their core. Indeed, in the case of the Eldonians, a third of the current residents returned from the municipal estates on the city's outskirts to which they had been 'decanted'. Again the contribution that these villages make to sustaining inner-city life and cultures cannot be exaggerated.

This is not to argue that issues of polarisation do not arise. Both the private and social housing developments, with their designs on 'defensible space' principles, are set apart to varying degrees in the built environment. The Eldonian and Athol villages clearly stand apart materially from the housing in their immediate vicinity (Peel, 1997).[13] What is needed is a sustained strategy for breaking down this isolation by encouraging the spread throughout the city-region of, to use David Harvey's (1997a and 1997b) terminology, the 'militant particularism' that produced those villages in the first place. But, as the MDC experience also shows, such a strategy will be expensive.

Between 1990 and 1996, the Vauxhall Housing Development cost the MDC nearly £25 million to develop the sites and to provide infrastructure, community support and training. The Housing Corporation provided £11 million and the private sector invested just under £10 million, bringing the total to £46 million (in historic prices; figures from KPMG, 1997b). All this for a development comprising two housing association 'urban villages' of 445 dwellings and a nursing home and 103 private flats and dwellings.[14]

The MDC's strategy for business development also raised some important issues, in terms of the balance between indigenous business development and the promotion of inward investment and, in relation to the latter, in terms of the potential for developing local employment initiatives. Of the 19,100 'gross' jobs and £548 million private sector investment claimed in total by the MDC, only 770 jobs and £21.6 million in investment (about 4% of the respective totals) have been directly produced by the 36 foreign-owned companies that have moved into the MDC's area. Most of this employment and investment was associated with eight major projects. Compare this inward investment from the private sector with its public sector counterpart in the shape of the new Customs and Excise VAT Headquarters in Queens Dock Liverpool and the offices of the Land Registry and Child Support Agency across the river in Birkenhead. Together, these three developments accounted for nearly £61 million of the Corporation's claimed 'private sector investment' (11% of the total), only matched in terms of scale by the investment associated with the Albert Dock complex, and about 2,000 jobs. Indigenous, relatively small-scale, business developments were thus clearly important in accounting for the bulk of jobs and investment, emphasising the importance of the less 'high-profile' programme of site development for managed workshops and grant support for small businesses and business associations.

What the experience of the relatively small number of inward investment projects (in both public and private sectors) did provide, however, was important examples of the feasibility of developing local employment initiatives around such projects – through the Costco Employment Initiative in North Liverpool and the Equal Opportunities Recruitment Programme of the Customs and Excise VAT Headquarters on the Liverpool Waterfront. Whilst the initial achievements of both of these initiatives were, as discussed earlier, relatively limited in terms of actual job outcomes, important lessons were learned in managing local employment initiatives, not least in terms of engaging local community groups and developing appropriate training programmes with training providers. The initiatives also underlined the basic point that local regeneration requires an active employment policy and cannot rely, as already argued, on now discredited notions of 'trickle down'.

Regeneration also needs a public-sector led strategy which avoids slipping into the kind of competitive 'boosterism' on which 'urban entrepreneurialism' is fundamentally based (Harvey, 1989c). The picture, however, is surely not as starkly black-and-white as Harvey makes out. Implicit in all the policies are tensions between out-and-out entrepreneurialism of the 'place marketing' kind and the pursuit of policies responding to issues of social need which, if not addressed, will result in there being ultimately no 'marketable place'. Thus the MDC was pulled towards community issues like social housing or training. Likewise, while participating in the Merseyside Image Campaign, Liverpool City Council is also developing a 'Poverty Strategy' and exploring positive discrimination policies like its Vocational Training Initiative (aimed at unemployed black youth).

What is important is politics, and the politics in question is one that can decisively weight policies towards the long-term development of democratically-decided and sustainable, welfare-based development. Given the instability engendered by the still parlous local economic situation, the development of such a politics is charged with difficulty. The current Labour administration of Liverpool City Council, for example, is beset by factional in-fighting as it attempts to get to grips with its serious financial difficulties. The danger is that this division will encourage an even more inward-looking politics.

To break out, on the one hand, of introspection and political inertia and, on the other, of the 'urban entrepreneurial treadmill', a coalition politics that builds on the strengths of 'locality' but transcends 'localism' is necessary. And there are signs of such a politics developing on Merseyside as the experience of partnership working develops – from City Challenge, to Single Regeneration Budget Partnerships to the 'Pathways Partnerships' established in the disadvantaged areas of the city-region targeted for special funding under the 'pathways to integration' priority of the Objective One Spending Programme.[15] There is clearly still some way to go in this 'building from the bottom' exercise in urban and regional development. The definition of 'social partners', for example, still excludes local trades unions.

There is also a need to ensure that the partnerships are genuinely empowering local community groups and are part of longer-term strategies to support and strengthen community-based social and economic development (Nevin and Shiner, 1995; Jewson and MacGregor, 1997a and b; and Mayo, 1996). But there are signs, not least because of the recent change in central government, that the previous era of contested governance on Merseyside is gradually evolving into one in which alternatives to 'trickle down' and unadulterated urban entrepreneurialism can be developed (Lloyd and Meegan, 1996). As the experience of the MDC appears to demonstrate, a simple focus on narrowly defined, property-led, regeneration is unsustainable. Urban regeneration, if it is to be truly effective, needs to be sensitive to the social, political, environmental and economic specificities of the localities in which it operates at the same time as being able to address the general processes which combine to shape them.

Notes

1. This geographical targeting of promotional activity recognises the different geo-

graphical scales at which 'image' formation can operate and 'image' perception can differ.

2. The exploits of the promoter of the event eventually became the subject of one of Radio 4's investigative journalism programmes, *Face the Facts* (broadcast in January 1998). According to the programme, the promoter had, over the past decade, been involved in a series of concerts that had been financial disasters. Of the four concerts investigated, two had been held in Development Corporation territories: the Jean Michel Jarre concert in London Docklands in 1988 and the Concert for Columbus four years later in Queens Dock, Liverpool. Urban 'spectacles' of this kind are clearly not guaranteed money spinners.

3. The Development Corporation also found itself criticised for giving some £50,000 to local groups to stage meetings and events critical of Christopher Columbus' role in history. Criticised for both its 'urban entrepreneurialism' and its community support, the 'Tall Ships' 'spectacle' was one it clearly could not win.

4. Indeed, the Corporation does appear to have had a marked public interventionist, planning orientation from the outset. Key appointments were made from the public and planning sectors. Its first Chief Executive and second chairman were recruited from New Town Development Corporations (Northampton and Warrington respectively) and many of the professional staff came from local authorities (with the Merseyside County Council being a major recruitment source for such staff as well as providing, as already noted, the Corporation's first Chairman).

5. ERM Economics (1997a and b) reviewed achievements in relation to six areas: business development, community and training, environment, housing, private sector investment and tourism. KPMG Management Consulting (1996, 1997a and b) carried out six major project appraisals of the Albert Dock, Atlantic Avenue, Brunswick Business Park, New Brighton, Vauxhall and Wirral Waterfront.

6. For those interested in such an exercise, the assessments of outputs and project appraisals carried out respectively by ERM Economics and KPMG Management Consulting provide a good starting point.

7. The figures quoted include final year figures provided by Chris Farrow, Chief Executive of the MDC prior to publication of the Corporation's final Annual Report and Accounts.

8. It should be emphasised that these comparisons are based on 1996 figures. The MDC would certainly argue that the Corporation's relative position would be significantly improved if the comparison were to be made in the year 2000, given the rising trend in its 'leverage' ratio and significant investments in the last five years of its operation. Chris Farrow, Chief Executive of the MDC, for example, felt strongly that the final figures of £668 million in private sector investment and 20,000 (gross) jobs would rise, respectively, to about £1 billion and 30,000 (interview).

9. The definition of the standard output measure for employment creation is 'Gross jobs' created by first time occupiers of newly completed development (new build and conversion) in the UDA measured in terms of full-time equivalents. All such jobs count whether they are in new firms or relocations but exclude jobs retained. '*Outputs . . . count against targets* on the date that an employer first takes up occupation of a new development, *regardless of whether the job is actually filled*'

(cited in ERM Economics, 1997a; my emphases).

10. The MDC had, for example, to reclaim ten times the area of land than its nearby counterpart in Trafford Park.

11. Even figures from the same source can be contradictory. Thus, for example, in the MDC's brochure *All Our Tomorrows*, an estimate of £321 million is given for the spending if the 61 million visitors that are estimated to have visited events and attractions supported by the MDC between 1984 and 1996 on the same page (p. 19) that estimated visitor spending for the Albert Dock alone has been £482 million.

12. In this context, it is interesting to note that the very last project grant cheque authorised by the MDC was for the Mersey Partnership. The cheque (for £1.6 million) was given with 'a strong steer' from the MDC Board that funding of such 'cultural pillars' as the museums should be prioritised in the organisation's future spending (interview).

13. On this point Chris Farrow, Chief Executive of the MDC, argued strongly that this apartness should be viewed less 'in a private income, private ownership – material sense' and more in a 'collective energy, collective organisation sense' (interview).

14. The public sector costs are also put into perspective when they are compared with the annual Housing Revenue Budget of Liverpool City Council. The budget for 1997/98 is just over £32 million – for the whole city. Compare this figure with the average annual public cost of the Vauxhall Housing Development – some £6 million.

15. The priority comprises measures 'put together as a coherent package of investments in the people of Merseyside, in particular young people, the long-term unemployed and others at a disadvantage in the labour market' (Commission of the European Communities, 1995, p. 107) some '. . . concentrated on a limited number of communities in the region, facing the worst problems' and complemented by 'action on the public transport system, to provide better access to work; and environmental works to improve the quality of life' (*ibid.*, p. 108).

Further reading

The essential reading matter published by the Merseyside Development Corporation is referenced in the text although, given the Corporation's demise along with its counterparts elsewhere, this may soon become difficult to obtain. While there is a wealth of historical material on Merseyside, there is a dearth of up-to-date writing on the area, especially relating to urban regeneration and the role of the Merseyside Development Corporation. This gap, at least as far as Liverpool is concerned, should be filled with the publication in 2000 of a set of volumes, provisionally titled *Liverpool Hearings*, by Liverpool University Press. These volumes will attempt to position the city – economically, politically, socially and culturally – and will include further reflection on the legacy of the Development Corporation's activities. In the meantime Tony Lane's *Liverpool: City of the Sea* (Liverpool University Press, 1997) is a valuable read, especially the last chapter 'Arrival and Departure' in which Tony speculates on how the city might and should 're-invent' itself.

4

Urban policy, modernisation, and the regeneration of Cardiff Bay

HUW THOMAS AND ROB IMRIE

Introduction

From April 1st 1987, Cardiff Bay Development Corporation (CBDC) was charged with the task of establishing 'Cardiff internationally as a superlative maritime city, which will stand comparison with any similar city in the world, enhancing the image and economic well-being of Cardiff and of Wales as a whole' (CBDC, 1988, p. 2). It will remain operational until April 1st 2000, two years later than the last of the English UDCs. CBDC always had ambitions beyond the physical regeneration of Cardiff Bay, which embraced the generation of new spaces of production and consumption which would mark Cardiff as a nationally and internationally competitive city. In this sense, regeneration strategies in Cardiff are connected to what Jewson and MacGregor (1997a, p. 5) have characterised as processes seeking to create 'a distinctive civic image' whereby 'the ambience and style of the city become economic assets'.

A central concern of this chapter is an exploration of the apparent vanity, even absurdity, of a provincial city of 300,000 population in seeking to compare and compete with maritime cities worldwide (with, for example, Boston, Sydney, Buenos Aires, even Venice). For Cardiff, read off most cities worldwide in a context in which city and place marketing have become centrepiece strategies. From Milan to Dublin and Cardiff, city boosterism has become increasingly characterised by what Jessop (1997, p. 28) terms 'shift in cities' roles on subjects, sites, and stakes in economic restructuring and securing structural competitiveness'. For Cardiff, its ambitious engagement in global competition is connected to the specific and distinctive political background of South Wales and what this has meant in terms of post-war state intervention at central and local level to physically restructure the city. As we shall argue, the city's development has assumed totemic significance in a project to modernise a regional economy which has been diagnosed as outdated and in decline.

Indeed, CBDC's proclamation, of grasping 'a unique opportunity for

Cardiff and for Wales', reflects the continuing political value of particular notions of modernisation. Far from CBDC simply representing an imposed central government solution to the problems of Cardiff docks, the Corporation is wedded to some of the more widely held political values within the regional politics of South Wales, and, as we shall show, the terms of CBDC's engagement has, in part, facilitated some continuation of local authority involvement in the redevelopment of the city. All of this might suggest that a form of corporatist planning has emerged in the city, yet the actions of CBDC are also linked to a 'capital logic', creating the 'right' type of physical environment for private investors, a logic which has created tensions in Cardiff, particularly with groups seeking to promote social and community regeneration. In this sense, CBDC is indicative of the wider context of urban policy in pursuing strategies which seem to represent segmented interests, yet claiming to be planning the city of the future for all.

In considering these themes, we begin by placing the origins of CBDC within the particular political ethos of urban modernisation and local corporatism. We then look at the role of CBDC in the future of the dockside community indicating its uncertain involvement in developing wide-ranging links with local community organisations. This section provides background for our subsequent discussion of the institutional linkages between CBDC and local actors and agencies. Here we highlight the role of political incorporation and consensus building as the basis for the political and community acquiescence to the wider development objectives of CBDC. The limitations of this process are displayed in a discussion of the plans for the exit of CBDC. We conclude the discussion by considering the future for the evolving institutional frameworks in Cardiff and generalise our findings to the wider context of urban policy in Britain.

Regeneration as modernisation

Rees and Lambert (1981) have argued that for decades South Wales politics has been dominated by the idea that it is an area in need of modernisation. This notion has been a central ideological support for a corporatist politics which has ensured that major initiatives, such as the Cardiff Bay regeneration, have enjoyed at least nominal support across the political spectrum, and from trade unions and employers organisations. Outright opposition has been limited to sporadic criticism from academics, residents groups or well-funded national organisations (such as the Royal Society for the Protection of Birds (RSPB), in relation to the Cardiff Bay barrage – of which more later).

Operationally, modernisation, in the Cardiff context, has three key features, all of which are found in Cardiff Bay. Its defining characteristic is economic; at the core of the political analysis so dominant in the region is the idea that the economy is undergoing a yet-to-be-completed structural change, a change which the state can expedite, largely through attracting 'cutting edge' employers. Over the last thirty to forty years, hopes have flitted from one kind of industry to another as the salvation of the South Wales economy, leading to shifts in the emphasis of government strategies. For example, in the 1960s and early 1970s, the manufacturing of consumer products (cars, 'white goods' etc.) were considered to be archetypically modern industries (and a variety of manufacturers of washing machines, televisions and other goods were enticed to South Wales by government grants). In the 1980s, mobile finance capital and 'high tech' firms were attracted e.g. the ill-fated INMOS located close to Newport, as were TSB and other finance companies. By the late 1980s, and early 1990s, consumption-led economic development strategies in South Wales (and elsewhere) viewed tourism and leisure as 'modern' industries, and heritage centres opened in the South Wales valleys.

Yet if the main thrust of promoting modernisation has involved economic development, the (re)development of the built environment has also had an aesthetic and symbolic importance (i.e. an importance over and above the functional necessity of creating a new economic landscape). This is the second feature of modernisation as a political project in Cardiff. As Hall (1998) suggests, an important part of contemporary urban regeneration involves the re-imaging of cities, creating new aesthetics of the place in order to cultivate new consumers of the emergent built environment. Likewise, the redevelopment of Cardiff's city centre in the 1970s and 1980s was, in part, a project of creating a capital city which looked the part (Evans and Thomas, 1988), and the £190 million Cardiff Bay barrage, which has aroused limited popular opposition and far greater environmental concerns, is the kind of major project which marks out a city as modern.

The final strand in the notion of modernisation, which has been powerful in South Wales politics, is of dynamism, change and progress. Peter Hain, now a Welsh Office minister, voiced it recently when talking of the great opportunity presented to Cardiff by the holding of an EU summit in the city in June 1998, which, he claimed, would

> be an unprecedented opportunity to showcase the emergence of Wales as a modern, dynamic society.
>
> (*South Wales Echo*, March 9th 1998)

The idea that an event in Cardiff can highlight the modernisation of Wales as a whole also has a long history. Indeed, we would argue that CBDC's 'mission' needs to be viewed as the latest episode in a political project which has, periodically, gripped the state at local and central level – the creation of a modern Cardiff which is a capital city worthy of national, and even international, recognition and which advertises the modernisation of Wales as a whole (Hague and Thomas, 1997). In the early twentieth century, a commercial bourgeoisie, finding its political feet, lobbied for Cardiff to be granted city status (and, in 1905, succeeded); it also undertook the development of a civic centre at Cathays Park which strove for grandeur and dignity, and ended up with strong colonial echoes in its formality, materials and design. In 1955, Cardiff became capital city of Wales, despite opposition from elsewhere in the Principality, and from around that time, the state centrally and locally struggled to devise a plan for the restructuring of the city centre which would do justice to its status. These struggles were reflected in a number of modernising projects, from the 1960s onwards, including a new shopping centre, the pedestrianisation of the main shopping area, Queen Street, and, most recently, with the city centre complete, the blueprint proposals for Cardiff Bay.

Of course, the massive city centre redevelopment proposals of the 1960s and 1970s reflected, and depended upon, flows of capital searching for profitable investment on a national, perhaps international, stage (see Cooke, 1980; Imrie and Thomas, 1993a for details). However, the particular forms which the proposals took, and justifications which they received, coupled with the close involvement of central government in supporting and guiding the nature of the city centre development, reflected the existence of a regional spatial coalition. Indeed, the particular politics of South Wales has been critical in providing support for Cardiff's spatial development. As Rees and Lambert (1981) note, a broadly supported view of the desirable trajectory for the South Wales economy, and Cardiff's role in it, emerged in the postwar period guiding local and regional state action and involving political relationships which were cemented with the binding power of the Welsh Office. Within this, Cardiff became identified as the place to spearhead the revival not only of the local, regional, economy, but to lead a Welsh economic renaissance.

The Cardiff Bay regeneration strategy, with its rhetoric about creating a 'superlative maritime city', and its sweeping away of allegedly redundant or under-used industrial space to allow for offices, leisure and other consumption-related uses, is, then, a contemporary restatement of the modernising theme of the 1960s, 1970s, and 1980s. Moreover, its proposal for a grandiose ceremonial mall, connecting the city centre to the waterfront (Bute Avenue), has clear architectural echoes of the early twentieth century civic centre (now an Outstanding Conservation Area), of bourgeois order and dignity becoming focal points for the appreciative gaze of the populace at large. Indeed, the vigorous place-marketing of CBDC reflects what Wilkinson (1992) terms an increasing consumerist style of urbanisation, emphasising new lifestyles, and the pleasure to be had from the self-conscious city landscapes. What then has CBDC sought to achieve and how well has it succeeded?

The development corporation's responsibilities cover an area of 2,700 acres, within which there were, at designation, 5,000 residents and 1,000 enterprises employing 15,000 people (Figure 4.1). It is one of the largest Urban Development Corporations (UDCs), with annual grant-in-aid in 1996/97 for example, of close to £60 million. Its 1988 *Regeneration Strategy* foresaw a physical and socio-economic transformation of Cardiff Bay with the creation of at least 3– 4 million square feet of offices, 5–6 million square feet of industrial space (including 'high tech' and 'modern business'), 6,000 houses (25% social housing), and a range of tourist and leisure facilities (CBDC, 1988). The strategy was to be flexible and market responsive, yet this in turn left the UDC vulnerable to volatility (and, especially, recession) in local property markets. In particular, implementing CBDC's strategy was de-railed by the property crash of 1989–92 and consequently, as Table 4.1 shows, CBDC no longer expects by the year 2000 to get very close to the

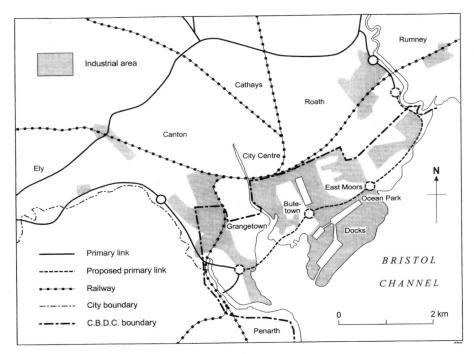

Figure 4.1 *Cardiff Bay in the context of the inner city*

Table 4.1 Proportion of target likely to be achieved by December 31st 1999, CBDC estimate

Variable	%
Private sector investment	68
Industrial floorspace	60
Office floorspace	46
Retail floorspace	66
Residential units	55
Land reclamation	65
Public open space	60
Permanent employment	52

Source: CBDC, 1996a

ten year targets it set in 1987/8 on a number of important variables (CBDC, 1996a).

While key targets will not be attained, there can be no doubting that CBDC has influenced changes in the spatial structure of Cardiff Bay and in popular and investor perceptions of the area (see figure 4.2). In particular, the corporation has pursued a modernising theme through encouraging development and the promotion of particular kinds of land uses in key sites. Often the changes have been uncontroversial. For example, in Ferry Road a mixed area of over 25 ha of derelict land, a refuse tip and a few scrapyards have been reclaimed, landscaped and developed as a retail park, with the prospect of some housing and a major commercial development in the near future. That site benefited from excellent sub-regional road links established before CBDC was set up; elsewhere, CBDC has implemented other improvements to the highway network which were planned by the (now extinct) South Glamorgan County Council.[1] As a result, road access to the land with the best prospects for high value development in the Bay – the so-called Inner

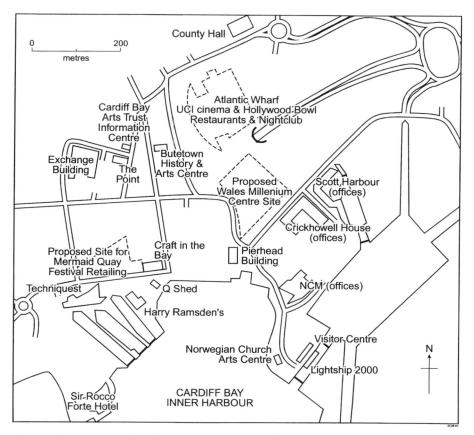

Figure 4.2 *The Inner Harbour showing key developments*

Harbour – has improved considerably in the last ten years.

A measure of CBDC's success in creating conditions for profitable invest-ment in this area is the influx of private leisure related developments in the last few years. For example, Harry Ramsden opened a restaurant in 1995, Rocco Forte is opening a luxury waterfront hotel in late 1998, UCI opened a 12 screen cinema complex in 1997, and a festival retailing complex will open on the waterfront in 1999. Taken with smaller developments, and the prospect of an Opera House/performance centre being opened around the millennium, then the corporation's talk of an 'arc of entertainment' being created in the Inner Harbour by the year 2000, and attracting two million visitors per annum, is not unreasonable. Yet, the regeneration of the Inner Harbour remains fragile and finely poised. CBDC has had to provide incen-tives of various kinds to secure some, though not all, of those developments. Details remain confidential, but its own documents refer to 'Investment incen-tives' in the Inner Harbour up to the year 2000 of close to £7 million (CBDC, 1997a, p. 52). In addition, there is (unattributable) information from offi-cers of land being conveyed at low or no cost for developments such as Harry Ramsden's fish restaurant.

While such developments are indicative of a regenerating local economy, they tell only part of the story. Also of significance in the revitalisation of Cardiff Bay has been an influx of relocating businesses from other parts of Cardiff, South Wales and beyond. The precise size of the displacement effect from elsewhere in Cardiff or its sub-region is unclear, and there are few exam-ples of very large offices moving from the centre to the Bay. However, an increase in vacant office space in the city centre has recently prompted county council concern that the Cardiff Bay area is proving attractive to local small and medium-sized office users, such as solicitors and accountants, if not very large companies, and certainly these are the apparent targets of the market-ing campaigns for recent developments.

CBDC's success, in its own terms, in sometimes generating precisely the kinds of physical and economic changes which symbolise a modernising soci-ety, must be set against its consistent failure to meet deadlines on three pro-jects which it has defined as vital to regeneration. Progress has been made on each, but it has been faltering. The biggest single project promoted by the corporation has been the controversial Cardiff Bay barrage, costing around £190 million, about 1km in length, and designed to create a fresh water lake of some 200 ha with a waterfront of 12km. From the very ear-liest days of the UDC's existence the barrage, and the associated waterfront, was promoted by CBDC as essential to successful regeneration. Its objective was to create the kind of waterfront setting that was extraordinarily fash-ionable among planners and property developers in the late 1980s following the apparent success of parts of London docklands and highly-hyped pro-jects such as Baltimore's waterfront. Waterfront development signified dynamism and, being modern, Cardiff Bay wanted it.

Unfortunately for CBDC, the barrage is technically and environmentally complex with implications, inter alia, for the diversion of sewers, ground-water levels (with apparent threats of flooding for some low-lying parts of the city), migrating fish, and wading birds who will lose important habitats under the impounded lake. An alliance of local residents worried about ris-ing groundwater and environmental groups (notably, the Royal Society for the Protection of Birds) campaigned against it; and though they were never likely to succeed (as local authorities and all but one local MP gave them no support), they did delay the necessary parliamentary approval of its con-struction for a number of years. Consequently, the supposed centrepiece of the regeneration strategy is being completed in 1998, some seven years or so later than anticipated.

More prolonged even than the barrage has been the delay in building Bute Avenue, designed as a grand boulevard linking the existing city centre with the 'vibrant' new waterfront developments in the inner harbour. The boule-vard is as important symbolically as it is functionally, for there already exists a direct route between the two areas and Bute Avenue will simply run in parallel with it. But the existing road, Bute Street, was the centre of the lively social and community life of the docklands heyday and, more, recently, Cardiff's red-light area (Evans *et al.*, 1984). More prosaically, it is fronted on one side by the poor-quality Butetown council housing estate, which does not present the kind of public face CBDC aspires to. Bute Street will remain, but will function as an access road to the estate; visitors will be guided on to the new Bute Avenue, where their views of the mundane reality of poor housing and above average unemployment will be obscured by heavy land-scaping.

Like the barrage, Bute Avenue's simplicity as an idea has belied the com-plexity involved in its implementation. The original plan was to wipe the slate clean – i.e. to clear existing firms in the industrial enclave of Collingdon Road and level an enormous railway embankment which runs the length of Bute Street. The new boulevard was to be developed in conjunction with a replacement light rail transit system and CBDC began acquiring premises in Collingdon Road in the late 1980s (see figure 4.3). It was the mid-1990s before the road was virtually empty; but the acquisition of premises was only a little more prolonged than the discussions between planners and urban designers from CBDC, the (then) City Council, and highway engineers from the (then) County Council over the details of the road's alignment and its connections with existing highways and urban development more generally at both ends. It is clear that the local authorities had a significant input into CBDC's final position on these matters, belying the idea that CBDC acts with complete autonomy and disregard for local institutions. Meanwhile, the £57 million price tag on the project had induced the Welsh Office to require CBDC to seek private sector finance through the Private Finance Initiative procedure.[2] A consortium including engineering firm Norwest Holst has suc-cessfully tendered for the project, but its bid does not include the new tran-sit system. With the announcement that the new Welsh Assembly is to be

Figure 4.3 *Over ten years on: industrial decline in the traditional industries of Cardiff Bay (Collingdon Road).*

located at one end of Bute Avenue, it is likely that the pace of development will now pick up, but it has taken a full decade to begin to get a start on site, with the associated uncertainty for residents and local businesses, and the loss of some manufacturing employment (Imrie and Thomas 1992; Imrie, Thomas, and Marshall, 1995).

A third major project has seemed, at times, to be even more ill-fated, namely the attempt to build an Opera House in the Inner Harbour. The details of what might fairly be termed a débâcle are straightforward. CBDC provided £2 million to fund feasibility studies and an architectural competition organised by the Cardiff Bay Opera House Trust as the basis for an application for National Lottery Funding (of over £50 million) as a millennium project. It also earmarked a prominent site for the proposed development. The competition attracted considerable public attention, and controversy surrounded the winning entry by architect Zaha Hadid. Meanwhile, working with the Welsh Rugby Union, South Glamorgan Council put forward its own proposal for a millennium project – a renewal of the national rugby stadium and its environs (in the city centre). Though the two projects were not, technically, mutually exclusive options, it was unlikely that a city of Cardiff's size would simultaneously gain funding for two such projects. In the event, the National Stadium bid was successful, while the Opera House bid was not; recriminations – some public, many private – followed (Crickhowell, 1997).

The Opera House bid fell in 1997, but, by 1998, a new proposal, for a Wales Millennium Centre, has secured Lottery backing and a start on site will be made soon. Nicholas Edwards, Chair of the Opera House Trust, has pointed out that the new proposal is uncannily similar to the final, desperately revised, version of the earlier bid, with the Opera House forming part of an arts complex catering for a wide variety of tastes. The project creates precisely the kind of image CBDC needs – 'a world-class showcase' in a 'landmark group of buildings' which will demonstrate the vitality, dynamism and modernity of Cardiff and of Wales (Wales Millennium Centre Project (WMCP, n.d).

These large schemes are only the more prominent examples of a concentration on property-led regeneration which has remained the consistent focus of CBDC's strategy. Table 4.2 illustrates the overwhelming significance of development projects in CBDC's expenditure.

However, one consequence of this major physical restructuring of Cardiff Bay has been the reduction of the supply of low cost premises suitable for manufacturing in favour of spaces for commercial and leisure uses. This, we would suggest, is one of the major reasons for the drastic reduction in the proportion of manufacturing jobs in the Cardiff Bay area since the mid-1980s, a period in which the significance of manufacturing employment in Cardiff as a whole has remained fairly stable (Figure 4.4). Construction jobs – as might be expected in a redevelopment area – have become increasingly significant, but are unlikely to remain at current levels for the medium-term (Figure 4.5). This economic restructuring is likely to have had a noticeable impact on the residents of Cardiff Bay. The 1981 census showed that in Cardiff Bay 35% of those who were economically active walked to work (compared to 15% in England and Wales as a whole) (Alden *et al.*, 1988). One might surmise that the concentration of manufacturing employment in the area, in the early 1980s, was important for a local population overwhelmingly in social classes 3b, 4 and 5. The loss of such employment would therefore be quite a blow. Certainly, unemployment in Cardiff Bay has not fallen relative to elsewhere in Cardiff or, indeed, in Wales. As Table 4.3 shows, the parliamentary constituency of Cardiff South and Penarth (which includes the Bay and its immediate hinterland) has risen in the unemploy-

Table 4.2 CBDC Expenditure 1987/88–1995/96

	£m
Development Projects	271.8
Social Projects	7.4
Financial Assistance to Business	128
Support Services	28.9

Note: figures rounded
Definitions are those of CBDC; 'Social Projects' includes £6.3m spent on development schemes (e.g. a school) and environmental improvements.
Source: CBDC, 1996b

Table 4.3 Parliamentary constituency unemployment by ranking consistent time series 1986–1995

Year	Cynon Valley	Merthyr Tydfil	Cardiff South and Penarth	Cardiff West	Swansea West
86	12	7	18	10	1
87	10	14	15	8	1
88	1	5	16	8	2
89	3	4	16	11	1
90	6	1	13	7	2
91	1	4	11	5	2
92	4	5	9	3	1
93	4	8	8	6	1
94	5	13	7	3	1
95	11	9	3	1	5

Note: The lower the number (i.e. ranking) the higher the unemployment rate.
Source: CBDC,1996a

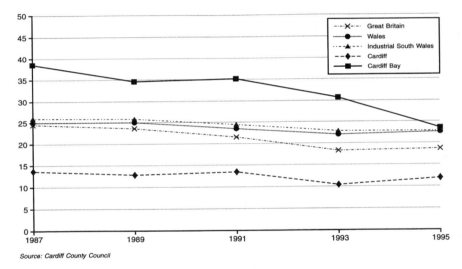

Source: Cardiff County Council

Figure 4.4 *Percentage employed in manufacturing industry, 1987–1995*

ment league table based on comparisons between Welsh constituencies.

On the other hand, CBDC's most recent figures suggest that some 1,300 people from South Cardiff/Penarth have gained employment in relocating firms formally assisted/advised by the corporation (of which over 900 have been in construction) (personal communication). These mixed outcomes of CBDC's decade of activity – some spectacular physical changes (see for example figures 4.6 and 4.7), but less apparent short or longer-term benefits for local people – have kept in political focus the simple, but awkward, question 'what does Cardiff Bay's regeneration mean for its original residents?' This concern

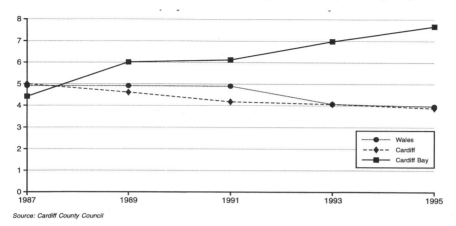

Source: Cardiff County Council

Figure 4.5 *Percentage employed in the construction industry 1987–1995*

Figure 4.6 *Private apartments in refurbished warehouse*

Figure 4.7 *New offices in the Inner Harbour*

about the distribution of benefits is particularly complex given the history of racism and stigmatisation directed at Butetown's residents. Some aspects of it will be explained in the next section of the chapter.

Cardiff Bay and local communities

In the background of local public and political attitudes towards CBDC is the complex role of Butetown – the 2,000 strong community geographically and symbolically the centre of the Bay – in the city's history and politics. Modern Cardiff owes its existence to the rapid growth of its docks in the late nineteenth and early twentieth century (Daunton, 1977). By 1913 it was the premier coal exporting port in the world, with a coal exchange, a sophisticated financial and service infrastructure, and a cosmopolitan dockland community unique in Wales. As the port declined in the interwar years, so the city's commercial centre of gravity shifted to coincide with its civic centre, around the castle and main shopping streets. Butetown was left, 'below the bridge', an area still socially and economically distinct from the city, and with little political clout as the economic significance of the docks declined and residential and commercial expansion was concentrated in the city centre and its northern periphery (Evans *et al.*, 1984). Memories of the institutionalised and personal racism which has isolated the dockland community throughout the twentieth century remain vivid (Sherwood, 1991). In the postwar period Butetown was the subject of municipal redevelopment, but otherwise the area's concerns appear to have been overshadowed by major peripheral development for private and public sector housing.

Meanwhile, the industrial areas around Butetown barely featured in local authority policy making, as planners concentrated on city centre redevelopment (Cooke, 1980). As late as the early 1980s, many significant industrial access roads in the docks were in private ownership with all that this implied for poor co-ordination and maintenance. The closure of the East Moors steelworks in 1978, with the direct loss of around 4,000 jobs, increased the political profile of the industrial enclave in the docks, and a great deal of public investment followed in land reclamation and highway improvement. However, the area was still conceived, in policy and political circles, as having an industrial future and, to that extent, it remained a place socially and economically distinct from the rest of a commercial city.

CBDC's strategy and rhetoric begin to challenge this relationship between the south and the remainder of the city, while seeking to establish a continuity between its activities and a particular reading of Cardiff's history. Its very name highlights tensions between these objectives. 'Cardiff Bay' is redolent of the city's maritime past, as are the names of new developments such as 'Atlantic Wharf' and 'Windsor Quay'. But until CBDC's formation, there was no Cardiff Bay. The provenance of the term lies in public relations not cartography or history, and a similar tale might be told of the other names which are gracing developments in 'Cardiff Bay'. The name marks a new beginning, a fresh start. as much as maritime continuity (Thomas and Imrie, 1993). Indeed, CBDC's approach emphasises a major weakness of modernisation to date – namely, the failure of 'modern' Cardiff to challenge racism and provide a fair share of jobs for Butetown's residents – and pins its hopes on the possibility of somehow 'drawing a line' and starting afresh in providing opportunities for local residents (Commission for Racial Equality, 1991; Thomas *et al.*, 1996).

We are told that Cardiff Bay, unlike Cardiff Docks, is going to be at the heart of a vibrant, internationally acclaimed maritime city (Pickup, 1988); a city which has turned its back on the waterfront is now to embrace it. In the absence of a local revolution in social attitudes, the 'respectable' citizens of Cardiff might not be expected to embrace easily the socially isolated, multi-racial Butetown, or the scrap-merchants, gypsy sites and heavy engineering of the docks. So it is the character of Cardiff Bay which will need to change if it is to become familiar territory to most inhabitants of the city, let alone a scene of superlative maritime developments. As mentioned above, there is evidence that the kinds of changes desired by CBDC are happening. Though there remain large areas of heavy industry and unused land, CBDC has invested heavily in physically transforming some prominent sites, especially in the Inner Harbour, the main tourist area. Private investment has been attracted to these areas, and new shops, hotels, cinemas and offices are built or under construction. New roads – generally planned prior to CBDC's arrival – have improved accessibility and the result appears to be an increasing number of visitors to the Inner Harbour. Parts of the Bay, at least, are being reclaimed by 'respectable' Cardiff.

Yet it is by no means clear that this process is effecting any fundamental change to local social relations and, in particular, the stigmatisation, largely through racialisation, of Butetown's residents. For instance, CBDC's housing programme and policies have more or less left untouched the deprived Butetown estates. As Rowley (1994) has documented, new housing developments and investment in Cardiff Bay have been concentrated in Windsor Quays, a waterfront site, Tyndall Field near the city centre, and Butetown. Whereas Windsor Quays and Tyndall Field have received significant investments in the form of new housing and major infrastructure works, Butetown has experienced small scale, piecemeal refurbishments. Rowley's (1994) research indicates that residents in Butetown were highly dissatisfied and that the refurbishments were seen as irrelevant to the needs existing on the estates. As Rowley (1994) shows

> while a bottle bank has been provided, broken elevators, faulty laundrettes, leaking roofs and windows remained. The CBDC has also failed to implement their suggestion that homes should be provided for Butetown's large ethnic population throughout the area.
>
> (Rowley, 1994, quoted in Hall, 1998, p. 151)

However, CBDC's so-called 'social projects' have delivered some benefits (a new school, a youth 'barn', and grants for projects) directly to local residents (see also Thomas and Imrie, 1993; Brownill *et al.*, 1996), though their funding has been a small proportion of CBDC's overall budget. Some local organisations and individuals have developed good relationships with CBDC's community development officers and have, perhaps, acquired the skills and aptitudes desired by those engaged in community 'capacity-building' (Brownill *et al.*, 1996). Organisations such as the Butetown History and Arts Centre are self-consciously engaged in securing a voice for residents' views of their own past, resisting any 'Disneyfication' associated with place-marketing. But these activities are put in a different light by the plan for Bute Avenue, discussed above. The symbolism of the road, its cordoning off of Butetown, is not lost on existing residents nor, presumably, will it be on visitors. What we see here is not a confronting and re-shaping of local social relations but a confirmation of them and an attempt to by-pass, or defuse, any of their consequences which might threaten CBDC's grand vision for the Bay.

Though individual officers may have acquired an understanding of the dynamics of the local community, for CBDC, corporately, it is evident that the local residents – particularly those of Butetown, so close to major development projects – represent an awkward and somewhat mysterious 'other', to be somehow negotiated with. Indicative of this are the comments in an interview in the mid-1990s of a senior planning officer who stated that in his view the role of the (few) community development officers employed by the corporation was to guide the planners through the community. Slightly more sophisticated are the comments of the current chief executive (in interview) who recognises that 'the community' does not exist as a single coherent entity or, in his words, 'that there's no structure within the community to have a relationship with'. His aim, nevertheless, is to have a relationship with residents, but its terms are significant: he has been working towards a 'contract', in which each side agrees not to ride roughshod over the other. In practice, this appears to amount to CBDC's being prepared, in principle, to deliver some community benefits in return for community acquiescence to its overall strategy, but finding itself, in practice, frustrated in trying to identify a grass roots body which can deliver the 'community's' side of the bargain.

But if incorporation of the residential 'community' has been frustrated by the inchoate politics (in the broadest sense) of Butetown, there are groups *within* CBDC who remain keen to secure long-term benefits for local people from the corporation's work. The small (3/4 person) community development team has worked hard for a number of years to distribute grants to local groups and to set up training schemes for local people. They have been successful in lobbying internally for the setting up of institutional defences of these activities for the post-CBDC period in the form of a Community Trust based on the example of London Docklands and Milton Keynes. That a need is perceived for such a mechanism, as opposed to entrusting the local authority with the task of continuing the community development effort, suggests that institutional tensions in the area are growing as the exit date approaches.

Institutional context and attitudes

We drew attention earlier to the regional corporatism which had presented Cardiff Bay as an innovation enjoying a broad base of political institutional support in Cardiff and South Wales more generally. But a consensus on the need to modernise the economy, and the spatial and social structure of a region, does not guarantee friction-free institutional working. Though CBDC has been a member of relevant networks and partnerships between local agencies engaged, for example, in place marketing, governance in the 1990s has also been characterised by competition between agencies for kudos and funding. In addition to these general tensions, there have been specifically local niggles.

We have noted elsewhere that the (then) city council was, in general, considerably cooler about CBDC's strategy then was the (then) county council, a difference related in part to the complexities of local Labour politics and in part to the financial interests of the city council, as landowner, in maintaining a buoyant city centre (Thomas and Imrie, 1993). The creation of the unitary Cardiff County Council in 1996 has simplified the institutional landscape of the city, but not necessarily made CBDC's life any easier. Within the majority Labour group (which holds 56 of the 67 seats) the dominant grouping is one associated with the old county council. Consequently, Cardiff County Council's urban/economic development policy has tended to follow the same boosterist path, with an emphasis on major projects, as did South Glamorgan. While this trajectory is supportive of the principles underlying CBDC's project and programme, there have been tensions arising from the lack of a strategic framework to guide and co-ordinate the work of CBDC and other agencies. The most visible illustration of these was the Cardiff Opera House episode. It has a two-fold significance for analysing local institutional relations.

First, it exposes institutional rivalries and the absence of a strategic policy or governance framework to curb them. The drawing up of competing bids for lottery funding was hugely expensive for the public agencies involved, and had important strategic planning implications in as much as both bids promoted mega-projects as vital components of the regeneration of parts of Cardiff which were functionally important for the sub-region (namely, the bus/railway station areas of the city centre and the newly redeveloped waterfront). Second, it underscores the extent to which the regeneration of Cardiff Bay had slipped down the priorities of the (Conservative run) Welsh Office by the mid-1990s. Though Lottery funding is independent of government, it is very difficult to imagine a CBDC-sponsored project even finding itself having to compete in the late 1980s and early 1990s. Cooperation, rather than competition, has been the guiding principle of the Welsh Office, and, as Thomas (1992) has shown, Secretaries of State for Wales have not been averse to interfering to ensure cooperation between agencies involved in the regeneration of Cardiff. By the late 1990s cooperation is more grudging – even in relation to CBDC's exit strategy.

The fragility of specific examples of institutional cooperation in pursuit of corporatist objectives is further illustrated by an example which contrasts strongly with the Opera House episode. CBDC has been instrumental in setting up the Cardiff Bay Training and Employment Group (CBTEG) which, since the early 1990s, has brought together CBDC's community officer with middle tier officers in the Employment Service, County Council, the local TEC and the Wales TUC in project based work aimed at improving the take up of jobs by local residents. It has had some modest success – its own estimate is that 1,350 local people have secured jobs as a result of its efforts (but close to 1,000 of these are in construction). Yet, after five years of working the future of the group is extremely uncertain as there is no successor to CBDC in the lead role, especially if the spatial focus on Cardiff Bay, or south Cardiff more generally, remains. No agency is willing to give priority to taking on CBDC's work and, more particularly, the county council is evidently distancing itself from CBDC's focus on the Bay area, while continuing to support the general ethos of city boosterism. CBTEG's difficulties, therefore, illustrate some general issues surrounding CBDC's exit.

At the time of writing, research on the progress of CBDC's exit strategy is heavily reliant on interviews with key personnel, some of whom have asked for non-attribution. As a result, checking some matters of detail is impossible. It is clear, however, that there is considerable tension between CBDC and Cardiff County Council over almost every aspect of the exit, including its management. Though CBDC is scheduled to be wound up by April 1st 2000, discussions about the management of the exit had barely begun in the spring of 1998. A Liaison Group appears to have been established to plan and manage the exit, though conflicting reports are received even about that (and no minutes are publicly available to date). It consists of a handful of leading councillors from Cardiff County and the Vale of Glamorgan Borough Council, together with the chair and vice-chair of CBDC. There are rumours of bitter disputes over the chairing of the group, but its very composition demonstrates the extent to which the development corporation has failed to become the forerunner of a more plural, or complex, network of governance institutions.

Neither representatives of local community groups nor businesses are included. It would be naive to conclude that some of these do not exercise some influence, particularly as the local authorities are as committed to 'levering in' private money as was CBDC, but it is clear that the intention of local government is to underline its importance as the key agency in a local corporatism: it is *through* and with local government that urban policy will continue to be delivered in this part of South Wales. This is wholly consistent with the Welsh Office view of urban policy delivery, which – with the possible exception of its setting up of CBDC – has never questioned the role of local government in the way in which it has been done in England, and which, in its latest guidance, has re-emphasised the importance of local authority discretion over spending (Welsh Office, 1997).

The implications of this for Butetown's residents are serious, for there is little evidence of a sea-change in local political priorities, and the welfare of Butetown's residents will receive no greater (and perhaps less) weight than, for example, the welfare of people in peripheral housing estates. Moreover, as mentioned earlier, in 1997, CBDC established a steering group of interested bodies with a view to establishing a Community Trust in the spring of 1998. The purpose of the Trust is to:

- enable social and community regeneration to continue beyond the millennium
- continue the provision of grant aid to local groups, and
- support local employment and training initiatives.

(Cardiff Bay Community Trust, 1998)

It is reasonable to interpret these objectives as a list of community-related activities commonly regarded as 'at risk' once CBDC winds up; the corporation, itself, proposes to donate £2 million to the trust and funding has been sought from the National Lotteries Charity Board, European Regional Development Fund and the private sector. Significantly, the county council appears not to be involved in the trust's funding and, indeed, is barely mentioned in its business plan. The trust represents an institutional defence of some of the community-related urban policy initiatives of CBDC, minor though these have been in terms of the corporation's overall funding. In constructing the defence, CBDC will be attempting to exploit its own networks within the local residential areas and local businesses, and among institutions.

Yet despite these arrangements, previous research has unearthed discontent and alienation amongst segments of the local community, especially among small firms affected by CBDC's plans (Imrie and Thomas, 1992). Such discontent helps explain the emergence, in early 1992, of the Cardiff Bay Business Forum (CBBF) as a voice of local small firms. The Forum has displayed a political astuteness not previously apparent among the Bay's smaller firms, and has been at pains to present itself as a responsible body, able and willing to enter constructive discussions. It is generally supportive of CBDC's strategy but concerned that local small firms are not yet seeing any benefits from it, while the inevitable disruption hits them harder than most. In particular the CBBF attracts a wide range of businesses, and particularly the very small ones (Imrie, Thomas and Marshall, 1995). Some of these, such as small shop and snack bars, have previously had no obvious vehicle for representing their interests, while the small engineering establishments or professional offices are evidently unhappy at the lack of vigour with which their interests were pursued (if at all) by the formal liaison committee with businesses set up by CBDC (membership by invitation) and the Chamber of Commerce and Industry, which was one of its members. There are indications that the CBBF is gaining acceptance as a legitimate voice of a stratum of the Bay's businesses but whether it exercises any influence over the Development Corporation remains to be seen; to date, there is little evidence of it, its chairman has argued that it remains marginal, and its very existence suggests that, smaller firms in the area remain unconvinced of the benefits to them of CBDC's strategy for renewal (Open University, 1997).

While agencies like CBBF and Cardiff and Vale Enterprise Agency (CAVE) are influencing, at the margins, aspects of CBDC's strategy, the most significant influence is central government (i.e. the Welsh Office). Though all UDCs have had their expenditures closely monitored and controlled, planning consultant David Walton (1990, p. 1) – who has worked with a number of UDCs – has commented about Cardiff Bay that 'central government . . . was more closely involved than would be normal elsewhere in Britain'. Indeed, a central and controversial feature of the regeneration strategy – namely, the massive barrage across Cardiff Bay – 'was in place from the outset' (*ibid.*). It is clear that, in Cardiff, central government involvement has extended beyond financial regulation to directly influencing a major element of the regeneration strategy and (we would speculate) more subtle influence over the overall nature of the regeneration package. However, perhaps this influence should not be conceived as an imposition from above, but more as a particularly important voice in a discussion and debate, in local and central government, about the future of the docklands which, as we have set out earlier, has been under way since the late 1970s. The Welsh Office, in this reading, is a regional actor, responsive to regional circumstances, rather than simply a transmission mechanism for national (UK) policies. For example, from 1995/96 the Welsh Office has set CBDC an annual target for 'local people into jobs' (currently 300 p.a.). This does not reflect a change in national (UK) policy, but simply more sensitivity – on William Hague's replacing John Redwood as Secretary of State for Wales – to those local/regional voices which had long been concerned about more obvious benefits being supplied to local residents.

Yet, there are a number of significant instances in which the fiscal and political controls of the Welsh Office are undermining the development of a city-wide corporatism in the city's spatial development. This is best illustrated in relation to CBDC's expenditure programmes concerning land acquisition, compensation to companies, and their utilisation of section 146 of the 1980 Local Government Planning and Land Act. In the latter instance, CBDC have utilised a provision in the Act which seems to state that if you have an organisation (e.g. a business) in the UDA that wishes to remain, then the UDC can sell property to you at a subsidised rate. However, in using this, CBDC has come into conflict with their original remit (making market returns on sales, etc.) and were told by the Welsh Office (via a Treasury directive) not to use the provision on expense grounds. Indeed, CBDC were reminded by the Welsh Office of their obligation to 'get the best practice possible' while exhorting CBDC to use the minimum assistance to achieve development. Nevertheless, CBDC have used the provision on one occasion but have concluded that it is 'too much hassle as we have to go to the Treasury in London each time for approval'.

Conclusions

An interim evaluation of CBDC in the early 1990s portrayed it as an organisation pursuing objectives largely shared by powerful political interests in the city and the sub-region (Thomas and Imrie, 1993). Its very existence almost inevitably caused friction with the local authorities, but hardly amounted to the kind of brusque assault on local government found in England. However, it was vulnerable to the charge of having 'missed the boat', for its reliance on property-led regeneration had left it becalmed in the property recession of the early 1990s and little had been achieved in the boom years of the late 1980s. As it approaches the end of its life, CBDC's mission and strategy remains, in essence, the same. However, the general economic buoyancy of the mid-1990s has fed through to a degree of investor confidence in at least some parts of Cardiff Bay and some of the spatial restructuring may turn out to be very significant. It can be argued that CBDC has managed the kind of expansion of Cardiff's commercial core into the Inner Harbour which is essential if the city is to have facilities of a range and standard necessary to be considered seriously as a regional capital in Europe. But, of course, if the city (and region's) quest for growth is unsuccessful, then it may have neither the population nor employment to support twin-centres.

Moreover, while there may be general agreement between CBDC and local political and business élites about the desirability of its strategy, question marks have been raised about its competence in pursuing it: for example, the famous barrage will be built, but was mired in parliamentary arguments for many years; and Bute Avenue, for which dozens of small businesses were relocated, remains unbuilt, and is now unlikely to incorporate the light rapid transit system that was promised. Moreover, the fiasco over the proposed Welsh Opera House exposed a deep-rooted antipathy to particular aesthetic and cultural modes of production and consumption, a politics not necessarily congruent with the wider modernising ethos of CBDC. The strategies of CBDC have also been implicated in generating spaces of social exclusion and a range of examples highlight the partial nature of the UDC's policies and practices.

Such partiality can be demonstrated in many different ways, from the absence of strategy to address the poverty in Butetown, to the poor treatment of small firms by CBDC in seeking to clear strategic sites such as Collingdon Road (Imrie and Thomas, 1992). Moreover, the low spend on community projects, set against infrastructure and road expenditure, leads to questions concerning who gains and loses from the policy practices which have emerged. For Harloe and Fainstein (1992), for instance, most of the changes in places like London and New York benefited the financial and producer service industries, higher-level consumer services and the new service class working within regeneration areas. Real estate and other property interests did well too. Cardiff Bay has displayed similar patterns of change while providing limited job opportunities to the poorer residents in the area. As this chapter indicates, short-term construction-based employment has been one of the dominant trends providing little by way of secure employment opportunities.

CBDC's policy trajectories have been implicated in these emergent trends yet one cannot claim that the UDC has introduced new objectives or approaches to urban policy in Cardiff. As we have argued, CBDC's strategies were closely tied to previous rounds of policy and were implicated in a deeply organised political commitment to (forms of) modernisation. However, this does not mean that when it leaves the scene the politics of planning and development in the city will be, of necessity, unchanged. For an unanticipated legacy of CBDC has been the resurrection and consolidation of an interest in the planning of the docklands among businesses and residents, and the creation of organisations such as the Cardiff Bay Business Forum, the Cardiff Bay Community Trust, and community organisations such as the Butetown History and Arts Centre which may well provide the kind of institutional infrastructure for sustaining discussion, debate and mobilisation in the future.

Acknowledgements

We gratefully acknowledge the assistance of two Economic and Social Research Council grants, Nos. R000 233525 and L311 253060, in supporting some of the research in this chapter.

Notes

1. In April 1996 local government in Wales was reorganised, with unitary authorities replacing the two tiers of counties and districts. Cardiff County Council – based largely on the boundaries of the old Cardiff City Council – has replaced Cardiff City Council and South Glamorgan County Council.
2. Private Finance Initiative (PFI) is a set of policies promoted by central government since 1992 which aims to 'Increase the flow of capital projects against a background of restraint on public expenditure. The public sector is encouraged to bring the private-sector more centrally into the operation of capital assets. It is aimed at harnessing private sector management skills, and at a transfer of risk away from the public sector, onto the private sector' (RICS, 1995, p. 7). It is these objectives which defines PFI, with a key underlying principle being 'that although the government may need to be responsible for the delivery of a particular service, or with the capital expenditure associated with providing it, the government does not necessarily have to be responsible for managing the service, or for undertaking the investment itself.' (*ibid.*, p. 9).

Further reading

Cardiff was catapulted from insignificance into being the largest town in Wales during the late nineteenth century and an understanding of the modern city is assisted by an appreciation of this period. Daunton (1977) provides an authoritative account of urban development from 1870 to 1914, but unfortunately there is no equivalent history of the years since. There have, however, been some useful city profiles, which provide overviews of recent planning policy (Thomas, 1989; Alden and Essex, 1999), and analyses of aspects of the politics of recent urban development (e.g. Imrie and Thomas, 1995; Imrie and Raco, 1999; Thomas, 1999) and economic development policy (Valler, 1996).

5

Tyne and Wear UDC – turning the uses inside out: active deindustrialisation and its consequences

DAVID BYRNE

Introduction

> ... it is the distinction of this river that it is man-made. Literally, its tidal flow, no less than its busy banks, its capacious docks, its magnificent piers ... are the handiwork of a generation of Tynesiders who have snatched a port from the North Sea and converted what was little better than a ditch into a great river. It is but fact to say that the Tyne of today is so vastly different from the Tyne of our grandfathers that its transformation offers one of the finest examples of applied local effort. (R.W. Johnson, 1925, p. 5)

The Tyne and Wear Development Corporation (TWDC) is now nearing the end of its life. It is possible to attempt an evaluation of what it has done, measured against a set of three benchmarks. The first is what this UDC was supposed to do, given the general brief from central government. The second is what it has done as compared with what it said it would do. The third, the only one that really matters in the long run, is what are the consequences of its actions for the people of Tyneside and Wearside. In evaluating that we need to pay attention not only to what the UDC has done, but how it has done it – to process as well as to outcome.

The operations of the TWDC are quite distinctive from those of the other estuarine UDCs. The Tyne and the Wear are different. These were and are rivers on which things were and are made, things (ships, topside engineering on rigs, etc.) which have to be made next to deep water because they are structures which go to sea. The Tyne is a port, and marine transport is important in relation to the operation of TWDC, but it is marine manufacturing which matters the most because marine manufacturing is the industrial activity around which the modern cities of Tyneside and Sunderland were made. Indeed the only way to define Tyneside as a place is precisely as riverside and it is quite conventional to use the same form, Wearside, despite Sunderland's existence as a distinctive local authority.

To understand the operations and effects of the UDC we have to understand both what went before it and the context in which it is operating. The first is a matter of reviewing the history of the making of the modern Tyne and Wear. On the Tyne we need to note that national ports policy after 1965, part of a national programme of modernisation, delocalised control over the river around which the city was constructed. On the Wear nationalisation drew shipyard sites into public ownership and made shipyard futures a matter of political manipulation rather than market determination. The second requires a brief identification of the short-term, supposedly market orientated, strategies, derived from a combination of new right ideology and the specific interests of property developing capital, which have underpinned urban development policy since 1979.

This chapter begins with a brief history of the modern Tyne and Wear and continues with an account of the operations of TWDC since its designation (set in relation to national urban policy) which will pay most attention to its anti-industrial development culture and the conflict between this and the two rivers' 'Civic Culture'. The emphasis here is on the cultural and political dimensions of a dispute between industrial production and financial circulation through land development. The operations of TWDC might have had great significance for differentiated reproduction/consumption, although in practice they haven't had much, and issues here will not be ignored, but the dispute about economic base is fundamental. This review will include an examination of the exit strategy of the TWDC and of its successors in terms of the form of City Challenge, Single Regeneration Budget, and Lottery/Millenium planning developments in and adjacent to its territory.

The development of the Tyne and the Wear – the making of two industrial cities

The Tyne and the Wear were both important medieval ports but their present form is the result of the industrial revolution in one of its first locations. The development of the Wear began in the early eighteenth century with the establishment of the River Wear Commission in 1717. The history is given in Milburn and Miller's (1988) account which has the appropriate and accurate title of *River, Town and People*. For the next two hundred years a series of river improvements and harbour works were carried out which culminated by the First World War in a river and harbour containing a good deal of coal and general dock facilities but primarily orientated towards shipbuilding. As Milburn and Miller note (1988, p. 7), the Newcastle journalist Duncan had remarked one hundred years earlier:

> Whether it was the River Wear which made Sunderland or Sunderland that made the River Wear may be difficult to decide.

On the Tyne, developments were complicated by the enormous historical powers of the City and County of Newcastle-upon-Tyne. It was not until the middle of the nineteenth century that a coalition of industrial capitalists and large local landowners, led by the radical newspaper proprieter Joseph Cowen and including the Duke of Northumberland who had very large holdings in and around North Shields, succeeded in establishing the Tyne Improvement Commission (TIC) which remade the river in the way described in the epigraph to this chapter. The TIC was a local body which was run by representatives of both local marine capital and elected local government. Its operations made modern Tyneside, a conurbation the population of which tripled between 1851 and 1911 on the basis of the enormous amount of marine related employment in shipbuilding, port activities and merchant shipping. Together with coalmining, also marine related through coal export, these industries provided the base of both the Tyne and the Wear until the early 1970s.

In 1968 the locally selected TIC was abolished and replaced by the Port of Tyne Authority (PTA), a centrally appointed quango. At the same time the Labour Government implemented a national ports strategy in which the Tees rather than Tyne was to be the key port for the North East. This is a particularly clear example of the way in which Labour's centralised efforts at national planning in the 1960s and 1970s facilitated Tory directed control of local affairs after 1979. Tyne port activities declined and were concentrated at the rivermouth leaving redundant quays in central Newcastle. After 1974 the now nationalised UK shipbuilding industry declined under the impact of global competition. After 1979 the Tory government was generally hostile to the heavy industries which were the traditional base of the organised working class. Beyond this general hostility it struck a deal with the European Commission about UK shipbuilding which permitted the provision of grant aid to Harland and Wolff's in Belfast on condition that shipbuilding was terminated on the River Wear. This politically mediated closure of the Sunderland yards is a crucial factor in relation to the operations of the TWDC.

The political innovation after 1974 was the establishment of the Tyne and Wear County Council with overall strategic planning responsibility through the medium of the Structure Plan. This constituted the planning regime within which TWDC was supposed to act, although in practice TWDC ignored it completely. The Structure Plan was the last local democratic statement of political priorities for land-use planning across the conurbation. The County Council had considerable difficulties with the Port of Tyne Authority as it had very little capacity to influence what the Port Authority would do, given the elimination of local authority representation from that body. There were a number of disputes, but their character was not indicative of the role that the Port Authority is now playing as a subordinate of TWDC. The basic position of the County was that it wanted the Port Authority to release some port land for general industrial uses, whereas the PTA insisted that any development of its land must be for uses which would generate significant port traffic, and hence port revenue. The Structure Plan did recognise that deep-water fronting land was a unique resource, declaring that:

> Land with a deep water frontage represents a special and scarce resource. Structure plan studies have established that there is sufficient land potentially available for general use without using this port owned land. In view of the limited amount of such land . . . land owned by the Port Authority in the following general locations should normally be reserved for port related uses . . . Whitehill Point (and others).
>
> (Policy ED4, 1979)

Essentially the County accepted the PTA's position as compatible with its general objective of maintaining industrial employment through industrial land development, although it would have liked more land released by the PTA for industrial estates. These sites were part of the overall county policy of designating strategic sites on which piecemeal development was to be discouraged. They were intended for large, job-creating, industrial employers.

The conflict between the County Council and the PTA is best interpreted as a dispute between corporatist labourism and marine capitalism. Corporatist labourism wanted to maximise employment. Marine capitalism wanted to maximise returns. These two objectives were not necessarily incompatible. Some port related developments, particularly in offshore construction, would be job-rich. Others, especially the development of ro-ro (roll on – roll off ferries) and associated transport clearing, would have limited job development potential. The issue and the dispute were real, but compared with what was to come this was a family squabble.

When the Tyne and Wear Structure Plan was being prepared the future of the Wear as an industrial river was taken for granted. At the time of nationalisation in 1977 there was no sense that the industry was in terminal decline. On the contrary there was massive investment in new covered shipbuilding facilities and in the development of new ship designs. In the late 1970s the industrial Wear was exactly that, and whilst it was recognised that shipbuilding employment was never going to return to the massive levels of the 1950s, the vision was of a modern and highly productive industry competing well in world markets and using its traditional river fronting sites to do so. Indeed at the end of the 1970s a new activity began on both rivers which was closely related to shipbuilding/ship repair and drew on the skill base of that sector. This was offshore engineering, and in particular on the Tyne the fitting out of the topside of rigs and other offshore oil structures with accommodation and production modules. This was a logical and appropriate development of the existing industrial complex.

Up to this point the developments accorded well with the traditional industrial/civic culture of the two estuarine conurbations, a culture based on highly paid male employment with little division between very highly skilled manual workers and supervisory/design grades in shipbuilding and merchant shipping. The maritime element here mattered a great deal. Tyne and

Wearsiders thought of themselves as a seafaring people – those who made ships and sailed them. Women played a part in this, not just as wives and mothers, but also as workers in fish processing and marine related industries like ropemaking. The planning intentions up to the late 1970s were meant to maintain the industrial complex on which this culture was based.

Enter the UDC

The premise on which TWDC has worked is that the material basis of this culture is finished for good. In his evidence to the House of Commons Select Committee on Employment, Balls, TWDC's Chief Executive, asserted that:

> industry within the river corridors is characterised by heavy marine-based manufacturing. Due to world market conditions causing decline in these sectors, there are also a growing number of derelict factories, warehouses, shipyards, slipways and dry docks along both rivers, with river or rail access primarily, many of which are unlikely ever to be used again for their original purpose. (House of Commons Select Committee on Employment, 1989, p. 309)

Indeed in *A Vision for the Future* (1990) TWDC went even further. Not only were the marine manufacturing sites derelict and 'unlikely ever again to be used for their original purpose', but the industrial culture created by the complex was holding back development:

> The economy of the North East has, until recently, depended on three industries: heavy engineering, coal mining and shipbuilding . . . For too long the need for a more diversified regional economy was not seen as important or necessary . . . Indeed the senior management of these three industries was so small relative to the numbers employed that the opportunities for aspiring talent were severely limited, so for the most part they left the region. The opportunities for local entrepreneurial activity, given the dominance of engineering, shipbuilding and coal in the market, were limited. With the decline of these three sectors, the banks of the Tyne and Wear, essential to the functioning of those industries, lapsed into dereliction. (TWDC, 1990, p. 4)

This is quite an extraordinary statement. Not only does it ignore the partnership corporatist strategy of industrial diversification which dates from the Special Areas initiatives of the 1930s, and which has generated enormous numbers of manufacturing jobs, albeit primarily in branch plants, but it caricatures the management structure and opportunities existing in the core regional structure. Even more importantly, it displays no sense of the way in which the county's core industrial structure has evolved through developments of the existing base in human capital and organisational knowledge. In particular it ignores the way in which shipbuilding, marine engineering and mining engineering have contributed to the development of offshore engineering, activity which is very important on the supposedly derelict banks of the Tyne.

The emphasis on derelict land is wholly in tune with national policy as contained in the legislation establishing UDCs and as expressed by various DoE ministers since. Ridley summed it all up in his evidence to the same House of Commons Select Committee on Employment (1989), when he appeared before them to tell them off for the critical tone of their original report. The identification of a conflict between land regeneration using the development industry, and the needs and wishes of local residents, is general in critical comment on the operations of UDCs (see Stoker, 1989, p. 161). What is distinctive about the Tyne and Wear situation is that subsidised 'catalytic' non-industrial development has been encouraged in opposition to the large scale existing industrial uses present on the rivers.

The planning strategy adopted by the TWDC was that of 'catalytic planning' defined thus by TWDC's expert planning witness at the public inquiry dealing with the East Newcastle Quayside, P.W. Jones, a director of Debenham, Tewson and Chinnocks, project advisers:

> There is, in my opinion, a distinction to be drawn between regeneration and redevelopment. Redevelopment of a site will succeed in bringing land and buildings into whatever use the market determines as the most appropriate for that site at that time. Regeneration on the other hand, aims to create new markets by increasing confidence and attracting inward investment. A regeneration project is needed to rekindle economic and cultural vitality of the site itself and also creates similar betterment to its immediate environs. When combined with other such schemes, it will also be a catalyst for sustained improvement and growth in the whole city and indeed the region. (Jones, 1989, p. 12, para 3.1.4)

In other words the task of the UDCs was to use public resources to get the market going, as TWDC put it (Jones, 1989, p. 16, para 3.2.2.4) to act as 'A Catalyst for Regeneration' – the catalytic image implies that the potential existed. It was only necessary to inject some energy into the system to initiate a self-sustaining reaction which would proceed without further intervention. This is not particularly good physical chemistry but it was clearly the sense in which the term was being employed. The use of the term 'flagship' by UDCs to describe particular developments is significant here. The 'flagships' are the physical representation of the catalytic process – the late twentieth century equivalent of Gray Street around the development of which the mid-nineteenth century urban renewal of Newcastle was hinged. The concept is quite well founded in the history of urban development and renewal but has only proceeded without any subsidy in urban centres and, very briefly, (Welfare Island where Olympia and York did succeed) on non-central sites in world cities. This was the process which was to be applied to derelict sites in clapped out North Eastern industrial towns.

TWDC has had four significant zones of operation (see Figure 5.1). The first is in West Newcastle on and around the site of the former Vickers factories, where it has supported the development of a non-contenious industrial park. The second is on and around Newcastle Quayside, and in particular the site downriver of the Tyne Bridge described as 'East Quayside'.

This is an interesting site. It consists of the upriver Newcastle Quays backed up by the original commercial centre of the City and a mix of warehouses and light industrial premises. Most of the original population has been displaced by the slum clearances of both the 1930s and the 1960s. There were a good many existing jobs of varied kinds, particularly in quite mucky light industries. The TWDC regarded this as a flagship site and has succeeded in attracting development to it. This site is the only one which has been subject to public inquiry, a public inquiry which seems to have taken place because the Chief Executive of the TWDC succeeded in infuriating a major transnational company, Proctor and Gamble, with regard to the future of significant premises owned by them within the area.

A Swiss based developer with connections with Gateshead's Jewish community proposed an alternative to TWDC's mix of offices, hotels, leisure outlets and housing, arguing for a continental style combination of residences and workshops for artists and craftspeople. However, TWDC did prevail and their style of development, unsustainable car parks included, has taken place. This area is now to be connected by a millennium bridge to the East Gateshead art park to be constructed around the former Baltic Flour Mill on the opposite side of the Tyne. This is a post-TWDC project originating with what some might characterise as the regional Arts mafia, but it is worth noting that the TWDC's developments made this, also massively publicly funded, exercise possible – a very clear illustration of the sedimentation of

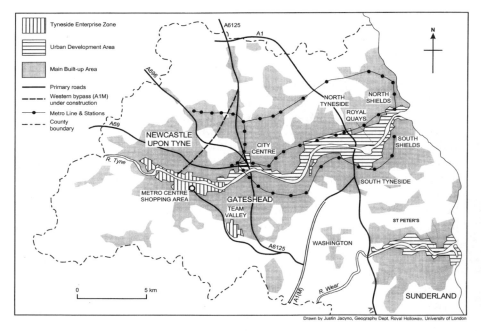

Figure 5.1 *Tyne and Wear Urban Development Plan*

recent history as the bedrock of future development.

East Quayside is an important case from an urban design point of view. The form of development achieved by TWDC with a very high office/car-park content is quite contrary to long-established county (but not Newcastle City) policies asserting the significance of public transport and the desirability of locating large office development adjacent to peripheral Metro stations. The developments are offices with a yuppie gloss in retail, hotel provision etc. The yuppification, especially in the Bridges area, is riding on the back of an established trend and the siting of Newcastle's spectacularly ugly new Law Courts on the quayside has facilitated the development of legal and related office interest.

The other two zones are the industrial Tyne and industrial Wear. On the Tyne the local Chamber of Commerce magazine identified: '. . . the TWDC's strategy (as one) of turning the traditional uses of the river inside out.' (1990, p. 25) Originally the Wear was identified as continuing in industrial use, a simple continuation of what was there and functioning. Indeed the Price Waterhouse original brief for TWDC stated:

> The Wear is very much a working river with many large industrial complexes located along its banks. . . . This area (the North Sunderland Industrial Belt) is part of the industrial heartland of the Wear Corridor, including two industrial estates, two shipyards and Wearmouth Colliery. It will continue to be dominated by heavy industry for the foreseeable future and the role of the TWDC will be to take positive steps to ensure that the industrial character of the segment is maintained. (1989, pp. 16–17)

As we shall see with the politically induced closure of the shipyards, the TWDC came in like a vulture to the sites and has turned the traditional use of that river inside out. Indeed, it has done so with considerably more success than it has achieved on the Tyne.

TWDC has been the planning authority for the whole of the North Bank of the Tyne to the rivermouth, for all the south bank excluding Gateshead and for all of the industrial Wear. It has had a series of proposals for development, beginning with those contained in its pre-designation brief. Although details, emphasis and, particularly, amount of action changed repeatedly, clear principles can be established. Basically TWDC has sought to insert non-industrial uses into previously industrial areas. The main argument for this has been that such uses will maximise land value returns but TWDC's Chief Executive has been quite explicit that this process is also one of cultural re-education. Land has been released for industrial uses, although the emphasis has been on B1 style business parks. However, with the exception of the Walker Offshore Park there was no specification of maritime industrial uses.

TWDC's first development was the isolated marina centred on St Peter's Village in Walker, which is a London Docklands style yuppie development on a former shipyard (Wigham Richardson's) site. Its 'flagship' site on the industrial north bank is the Whitehill Point land in North Shields, renamed

Figure 5.2 *St Peters*

Figure 5.3 *Royal Quays, Tyne and Wear DC*

for marketing reasons as Royal Quays. South of the Tyne the UDC has done much less. It supported the existing Hebburn Village development, initiated by South Tyneside Council, which again took former shipyard land, but not, as was originally intended, the Hebburn drydock which is the largest such facility in the UK. Some expensive housing has been built and a hotel is being constructed on the former Velva liquids site (renamed 'Littlehaven') at the South Shields rivermouth. The hotel has been the source of significant local controversy of which more subsequently. Very late in its life the TWDC acquired a small shipyard adjacent to South Shields Town Centre through a deal with its private owners, although the company had been previously threatened with compulsory purchase. This is being developed as another St Peter's style scheme, but this time largely composed of social housing. Otherwise TWDC has left the PTA with Tyne Dock as the centre of port operations. The offshore area between central South Shields and Tyne Dock and on the North Side of the Tyne has been left alone.

In Sunderland TWDC's original contribution was to demand residential land values for a crucial site adjacent to the one of Sunderland Ship-builders yards. This land, originally belonging to the local authority and intended for industrial development, had been taken over by TWDC. A Greek consortium, with local management and workforce backing, was seeking to acquire the site for a holding ship repair operation with a view to return-ing to shipbuilding in a world market context of a shortage of capacity for merchant shipbuilding. The enterprise did get going but is now not operating and Sunderland has only some limited ship repair sites left in action. The key Sunderland story is that of the development of St Peter's Riverside.

Whitehill Point – deindustrialisation in the face of change

The crucial site on the Tyne, both symbolically and in practical terms, is the port authority land at Whitehill Point and around Albert Edward Dock in North Shields. Here TWDC had the capacity to do the most damage on Tyneside and encountered the most developed and sustained opposition. This site consists of some 200 acres between Howden Road, which separates it from the 1930s slum clearance Meadowell Housing Estate, and the Tyne. It is centred around the still operating Albert Edward Dock and contains the ro-ro terminals for the Scandinavian passenger ferries together with other port facilities.

The PTA has always regarded this site as one of its prime assets. It resisted any non-port development proposals from Tyne and Wear County, and seems to have agreed a sale to TWDC only subsequent to Tynemouth's then Tory MP blocking a PTA private bill which sought to extend its development pow-ers. The TWDC could have vested the land but seems to have preferred a forced sale, under the terms of which TWDC obtained an option to purchase both the site and the Albert Edward Dock and associated quay frontage.

In late 1988 TWDC mounted a development competition based on an Invitation to Submit Proposals which stated:

> The Corporation's objective in inviting proposals is to achieve the rapid and successful regeneration of the site with high quality mixed use development, levering private investment and producing a satisfactory return to the public sector in terms of land price. Employment creation is also regarded as a major objective, and a target of 3,600 full time jobs is considered feasible (p. 3).

Three developers submitted schemes which were put on exhibition for public comment, although TWDC made it plain that it was picking a development consortium rather than a scheme – the schemes were illustrations rather than intentions. The project was awarded to the 'Royal Quays Development Consortium' whose original proposal was for 1,440 dwellings, 500,000 sq ft of business/industrial park, 390,000 sq ft of retail development and 390,000 sq ft for leisure related developments.

A vigorous opposition to these proposals was mounted by the North Shields Riverside Action Group. This was based around the Trades Council's TUC sponsored North Shields Peoples Centre and included supporters from community groups in central North Shields and on the Meadowell Estate. This large 1930s slum clearance estate was built as a replacement for the bankside slums which were North Shields' traditional sailor-town. Many of the estate's residents were merchant seamen, fishermen or ship-repair workers, and these industrial connections persisted throughout the post-war period. However, the estate was always the poorest part of North Shields and it has been the worst affected by deindustrialisation (see Byrne, 1989). Formal male unemployment rates are now of the order of 50% and only about a third of all households are connected with regular wage labour. The other elements in the coalition represented traditional trade unionism and, in the form of the North Shields Chamber of Trade, traditional shopping interests in central North Shields.

The Riverside Action Group produced an alternative planning brief emphasising marine industrial development, job creation and open riverside leisure access. It argued that the collapse of the Soviet empire (foreseeing this in 1989) would open up the Tyne's traditional trading connections with the Baltic and arctic Russia, offering the possibility of large scale new port trade. The moving southwards of North Sea oil exploration meant the possibility of service jobs. Finally, new offshore technologies, particularly wave-power, were identified as logical developments of the existing industrial structure and the basis for major industrial development after the year 2000. The group suggested that in the short to medium term the site could be used for port-development around Baltic trades and oil servicing together with light industry. The long-term potential would be developed around a wave-power research centre directed towards the development of production processes for this very promising technology. These proposals, including the leisure component, were compatible both with the Structure Plan and with detailed River

Tyne – Local Plan for Recreation and Amenity of 1983. The group, together with North Tyneside Council, drew up a scheme layout based on these proposals, which would produce some 4,200 permanent jobs with minimal displacement component.

TWDC proceeded to an outline planning application, made of course to itself, in July 1990. This repeated the earlier proposals with some modifications. North Tyneside MBC considered that the proposal constituted a major departure from the Structure Plan, not only in relation to the change of use of a strategic industrial site, but also because of detrimental impact on existing shopping centres, particularly North Shields. The proposal also contradicted existing housing land policies. The retail development would be almost exactly the same size as total retail space in North Shields Town Centre. In its formal response to TWDC North Tyneside Borough Council (1988) observed that:

> We are aware that the balance of the scheme has been determined to enable leverage ratios between public and private sector funding to be met. This, in our view, is an inappropriate way to plan the development of such a significant strategic site.

Despite the opposition of the local authority, and of the Tyne Port Users Group representing existing industrial capital, the Secretary of State did not submit the application to a public inquiry. TWDC was therefore able to award planning permission to itself. Development has proceeded but rather slowly. The whole future of the site was thrown into confusion by the Meadowell riot of 1992. Subsequent developments here have involved the coordination of initiatives on the Meadowell, funded in a variety of ways but primarily as part of a City Challenge scheme, with Royal Quays developments. Most immediately there has been a massive reduction in housing stock and densities on the Meadowell associated with the largely voluntary movement of much of its population to other areas in the wake of the riot. This movement reflected the availability of much better council housing throughout North Tyneside as an ageing population vacated more desirable estates. City Challenge is usually considered to be people-centred but in North Shields its operations have coincided with the removal of much of the existing local population and have taken the form of large scale land development which is not particularly relevant to the pre-existing local population. For example, despite rhetoric, there is no evidence that the Siemans factory (a perfectly acceptable branch plant development) has offered much employment to Meadowell-originating people. It draws from a wide catchment area. The overall effect has been to sanitise the immediate neighbourhood of Royal Quays and restore the possibility of some development (see Geddes, 1997 for a discussion of these developments in detail).

By late 1997 some developments on Royal Quays had taken place. A mix of retail (factory shop) and leisure uses with some light industrial development (one of the largest elements in which simply moved about a mile within North Shields) occupy about 300,000 square feet. TWDC anticipates some

560,000 square feet by the year 2000 which implies a very rapid increase in development, but even that total would be less than half of the original intended. There are 771 dwellings, many of them social housing, with 1,028 anticipated by the year 2000. The present level is about half that originally anticipated. Developments are well below target, despite part of the site being accorded Enterprise Zone status. All the development could have occurred on pretty well any site or set of sites within North Tyneside and there is no particular logic to the spatial coincidence of any of the elements.

Sunderland – the end of shipbuilding

Since the Sunderland yards were nationalised on closure their sites were transferred to the ownership of the TWDC. TWDC's first action was to demolish the new covered shipyard at Southwick, one of the most modern facilities in the world. TWDC figures have been quoted as saying that this was done in order to demonstrate that never again would shipbuilding be a significant source of employment in the town. The main figure opposing these developments was the then MP for Sunderland North, Bob Clay. Clay, remarkably, gave up being an MP to become an officer of a company established at Pallion in order to provide a base for a revival of shipbuilding when the EEC moratorium on it in Sunderland expired. Vital to any such project was the availability in Sunderland of a fitting out site at Manor Quay on the north side of the river below the bridges.

TWDC's operations in Sunderland have two important elements, although there have been various smaller scale tarting up activities going on in other places. The first begins with Hylton Riverside, an area of long redundant coal drops. Here a riverside business park has been developed. This is the western part of the 'Sunderland Enterprise Park' and includes a peripheral retail development of some 100,000 plus square feet, yet another nail in the coffin of retail activity in Sunderland town centre. Otherwise, in content the park is a reasonable land use although the employment offered is of a sort which has limited relevance to the residents of contiguous areas of overspill council estates, despite the development of dedicated training schemes. The eastern half of this area is the site of the former Southwick shipyard – demolished as exemplary action. Although this was bitterly resented, it was very much a *fait accompli*. Since the new B1 uses on this site can be represented as industrial, no change of use was involved and no planning consultation was required.

The same is not true of the below-bridge sites which comprise St Peters Riverside. What has been done here matters but even more interesting is the way in which it was done. Although the whole area has always been presented as an integrated whole in TWDC documentation, there has never been a planning proposal which dealt with it as a single unit. Instead it has been handled salami fashion, slice by slice, with a series of self-granted planning permissions allocated a bit at a time, despite the fact that they all contradicted the implications of the Tyne and Wear Structure Plan which was, nominally, the overall strategic guideline governing development.

Sunderland North Dock has become another marina style development, and some 415 dwellings have been constructed, again many of them being social housing. The most important development has been that of the new St Peter's Campus of the University of Sunderland which houses that establishment's schools of IT and Business. Whatever the merits of this expansion of the University of Sunderland, there was no need at all for it to be on this site. A number of others were generally more convenient for the existing University facilities. However, the building of this development has sterilised Manor Quay and ensured that the Wear will not again build ships. This contributed to the closure of Wearmouth Colliery in that the abandonment of the upper river forced all the dredging charges onto it and made it less viable than other pits at the time when it might have been possible to retain deep mining in it. The Colliery site has now been cleansed by TWDC as a football 'Stadium of Light'.

TWDC (1988, p. 7) have recently stated that: 'An early strategic decision involved switching the industrial and commercial focus of Sunderland toward the A19 in the west, and returning the downstream areas to living, learning, and leisure.' In fact far from being a strategic decision this seems to have been a result of the quango doing what it was told when the shipyards were closed. TWDC has not 'taken positive steps to ensure that the industrial character of (Sunderland) is maintained' (Price Waterhouse, 1989). It has done exactly the opposite.

Conclusion

> The fundamental message underlying the Statement is that Urban Regeneration is a 'rolling programme', a long term process of **reacting** [my emphasis] to the opportunity and new requirements that will always occur as the needs of the people and businesses in cities and urban areas evolve and change. To avoid returning to the scale of dereliction and the 'log jam' of inactivity initially faced by the corporation, continuous efforts and resources and a structured approach will need to be focused on regeneration activities in the future. (TWDC, 1988, p. 1)

TWDC has spent, on its own calculations, £408 million of public money to achieve £818 million of private sector investment. This is not a good performance. At a public private gearing ratio of 1:2 it is massively below the ratio of 1:3 regarded as the minimum acceptable in urban development.[1] It argues that it has 'created or safeguarded some 28,000 jobs' (TWDC, 1998). This seems to be the total employment in its territory and the claim is plainly absurd. At the end of 1992 the TWDC was claiming to have created, variously, 6,296 or 9,237 jobs. Robinson *et al.* (1993, p. 44) concluded that in reality the number of jobs created was much less than 1,000. Even taking the lower of TWDC's 1992 claims and being very generous about 'creation' there may have been some 4,000 jobs created by TWDC's interventions. That gives a price of more than £100,000 per job which is a very high unit cost indeed. As to safeguarding Robinson *et al.* show how this is a term applied to jobs existing on site or moving into these sites from elsewhere in the conurbation. This is pure displacement.

TWDC explicitly rejected a strategy for the regeneration of industrial Tyneside and Wearside which was based on the revival and futher development of a marine manufacturing and port-trade base. Instead it asserted a property orientated approach, but was actually singularly unsuccesful in bringing it to fruition. Its original proposals included a series of 'yuppie' exclusive villages separated from existing working class residential areas. Only St Peters Basin takes this form and it has been notably unsuccessful. The original yuppie development at Royal Quays, intended even to have its own primary school, is now in large part social housing. Indeed 25% of the 4,000 housing units developed in TWDC territory are social housing – all housing association with such high rents that they have become ghettos of the completely benefit dependent. The housing for sale is overwhelmingly rather basic and cheap standard units and is quite affordable by many employed working class people. These are not 'citadels of the rich' although that is what they would have had to have been to achieve the 1:3 gearing ratio intended. The retail schemes have been more damaging, not because they are particularly exclusionary in content (much of the development is factory shops selling seconds), but because they have exacerbated the serious damage done to town centre shopping in Tyne and Wear by the Metro Centre and other peripheral schemes.

The long term impacts of TWDC will be the deindustrialisation and sterilisation of the River Wear and the change in the processes of planning which its procedures involved. Healey (1992, p. 8), Professor of Town Planning at the University of Newcastle and a member of the TWDC Board, has summarised the general character of these processes and their outcomes rather well:

> The net result in the conurbation was a considerable flow of subsidy to particular kinds of activity, in a situation where local authority and regional assistance was being reduced. Public subsidy had thus switched from providing support for the demand for land and property in various ways to encouraging property supply. Much of the subsidy was spatially targeted, to inner city areas, but also to zones away from established centres for office and service activity. These locations thus looked set to alter established spatial patterns in the conurbation. The subsidy was accompanied by agencies urged to be helpful to the private sector, and encouraged to engage in energetic promotional activity. . . . Urban policy was thus directed at transforming the spatial structure and institutional relations of the conurbation to reflect post-industrial conceptions of urban structure and lifestyle.

In her interesting review of alleged post-modern planning in Newcastle, Wilkinson (1992, p. 178) remarked that:

The T. Dan Smith era was concerned primarily with civic pride and a utopian version of the city as an urban machine fit for living in. It was essentially a modernist vision with a strong social welfare component, managed by the public sector on Keynesian functional principles. (1992, p. 177)

She contrasts this with:

the post-modern city . . . characterised by a shift away from comprehensive redevelopment projects, characteristic of the 1960s and 1970s, towards the planning of urban fragments, evidenced in the mosaic effect created by the development of the new urban villages, flag-ship schemes, self-contained waterfront developments and cultural quarters. These islands of renewal also act as highly visible symbols of urban regeneration and, as such, they are regarded by public and private-sector agencies as vital ingredients in the place-marketing process.

It is not just that TWDC got it wrong, although it misjudged the property market and the kind of developments to which it was committed materialised on a far smaller scale than was intended. It is not just that proposals, by emphasising property-derived criteria of exclusivity, were actively against the interests of a number of poor riverside communities. It is not just that supposed catalytic planning has never got going with the whole development history of TWDC sites being one of massive direct, fiscal and land preparation subsidies to the private sector and an extraordinary representation of other quangos in actual developments achieved. TWDC has gone against the whole cultural bias of the county in which it is located. It continues to do so. At the time of writing it has just awarded itself planning permission for a hotel at 'Littlehaven' (like 'Royal Quays' a TWDC invented name) against sustained local objections. The organisers of the objection commented that:

Visitors obviously take more priority than people who have lived in South Tyneside all their lives . . . For just £4 million they have bought our heritage. (*Shields Gazette*, 14 November 1997)

None of this has been subject to any kind of democratic control. The process of public inquiry has been initiated only when the TWDC's Chief Executive was fool enough to seriously annoy a major transnational company. The democratically established Structure Plan was wholly ignored, easy enough to do when the democratic body which drew it up, the Metropolitan County, had been abolished by central government. There was no sense of any strategic vision. TWDC's only success in terms of benchmarks as described at the beginning of this paper, has been in terms of some development achieved on its sites – effectively any development that could be got. Development proposals have changed continually and often fundamentally. Those TWDC spokespeople who have been claiming for it a 'strategic vision' as its activities are handed over to English Partnerships must have a highly developed sense of irony. It has been an essentially reactive agency. Its only proactive strategy has been deliberate deindustrialisation.

Unfortunately it seems very likely that the evil that TWDC has done will live on after it. Not only has it sealed the future of crucial strategic sites but it has played an important part in the achievement of a culture of impotency and oligarchy in Tyne and Wear. Successor urban development proposals, including City Challenge and SRB schemes, seem as far removed from democratic accountability as anything done by the TWDC. There is of course consultation, as with the TWDC's panels but these panels have no representative foundation, no channel of wider accountability, and have never had the slightest determinant influence on the character of developments. They have been confined to consideration of details of implementation, a process repeated exactly with City Challenge schemes. We have had major planning without consultation, consideration or democratic determination. This style has even penetrated Gateshead, although the local authority has been seduced here. An originally reasonable proposal for an art facility in the interesting Baltic Flour Mill building costing some £10 million has become a series of major arts regeneration projects funded by lottery money at £80 million and involving the displacement of an existing working class community to facilitate their development. The style of the politics of planning has become exactly the post-modern particularism described by Wilkinson. There is no sense of a general, universalistic, strategic vision.

This is beautifully illustrated by the Millennium Fund financed 'International Centre for Life' on the site of the old Newcastle Cattle Market. This has been described as 'an expensive jobs club for former quango (TWDC) employees' (Cllr Kevin Jones, *The Sunday Sun*, 18 January 1998). It will be headed up by the former TWDC Chief Executive and combines a 'bio-science centre' (biological science park), 'Helix: a themed indoor family entertainment centre and activity centre complete with a "river of life dark walk" and simulator ride', and Newcastle University's Clinical Genetics Research Institute. The Science Park/Academic element is fine but the city centre location for it is not a good one and the cost of £54 million is massively greater than what those elements would have cost on a more appropriate site.

TWDC's *Regeneration Statement* (1988) constitutes its withdrawal strategy. In practical terms this is a list of sites on or adjacent to TWDC territory with some proposals for what might be done with them in order to attract investment. In principle it is an assertion of anti-industrialism and of a politics based on reaction to forces which are considered to be beyond the control of any local civic democracy. The term globalisation is not used, but the overweening power of global capital is implicit as a background factor. The signs are that this will be the nature of regeneration politics under Blair, a man who believes absolutely and explicitly in the power of the global.

TWDC is a classic illustration of the failures, not of free market capitalism with which it has little connection, but of anti-democratic central direction conducted without reference to the political culture of the place concerned. If there is any lesson to learn from the experience, it is probably that Derek Senior got it absolutely right in his minority report of the Redcliffe

Maude Commision on the reform of Local Government and Peacock and Crowther-Hunt got in right in their minority report of the Royal Commission on the Constitution. The government and planning of development in this country requires effective democratic processes which can handle whole cities and whole regions. The last thing required was colonial administrations in the form of UDCs.

Note

1. Indeed the performance is worse than this. Private sector investment seems to include that made by housing associations and by the University of Sunderland. These are public sector bodies and as 25% of the housing in TWDC's area has been provided by housing associations and the St Peter's Campus is one of the largest developments in TWDC's territory, the real public private ratio is much less than 1:2.

Further reading

Byrne, D.S. (1989) *Beyond the Inner City*, Milton Keynes: Open University Press, especially Chapter 5, provides an account of the development of Tyneside in the post-war years with particular reference to the history of planning in riverfront North Shields. Also, see Byrne (1987).

Mess, H.A. (1928) *Industrial Tyneside*, London: Ernest Bess is the benchmark for any study of the industrial conurbation – an account of its social and economic conditions as it passed from growth to decline with extraordinary rapidity.

Colls, R. and Lancaster, W. (eds) (1992) *Geordies – Roots of Regionalism*, Edinburgh: Edinburgh University Press: the various chapters address exactly the character of the distinctive industrially derived popular culture of the North East of England.

O'Toole, M. (1996) *Regulation Theory and the British State: the Case of the Urban Development Corporations*, Aldershot: Avebury: pays particular attention to the Tyne and Wear UDC, while Geddes, M. (1997) *Partnership against Poverty and Exclusion*, Bristol: The Polity Press includes a case study of City Challenge in North Shields which illustrates the combined effects of UDC and City Challenge with particular regard to political process.

6

'Good Conservative policies translated into practice': the case of the Teesside Development Corporation

FRED ROBINSON, KEITH SHAW AND MARTY LAWRENCE

Ten years ago, Margaret Thatcher walked in the wilderness of derelict Teesside. Yesterday she was back, walking almost hand in hand with John Major, to trumpet the transformation of the Teesdale development in Stockton. Spanking red-brick office blocks with ornate ironwork have sprung up beside the Tees . . . 'It's good to be back to see all this' said Baroness Thatcher as she stood in front of Dunedin House, the Headquarters of the Teesside Development Corporation. 'Good Conservative policies translated into practice'. (*The Northern Echo*, 17 April 1997)

Creating a new image for Teesside was the priority for the Development Corporation. There was a need to break from the past, and to be at the forefront of environmental change, improvement and rehabilitation. There is no doubt that Teesside Development Corporation approached the task with enthusiasm and skill. It felt the need to be bold and imaginative, and it is not surprising that, in the process, some feathers were ruffled (Angela Eagle MP, Under-Secretary of State for the Environment, Transport and the Regions, HC Hansard, 14 January 1998, col. 464).

Introduction

No one could accuse the Teesside Development Corporation (TDC) of 'going gentle into that good night'. A few weeks before its official demise, Teesside Development Corporation (TDC) was singled out in a House of Commons debate not only for its failure to make adequate exit arrangements with the local authorities but also for allegedly shredding papers relating to its activities. In his wide ranging criticism of the Corporation, Ashok Kumar, MP for Middlesbrough South and Cleveland East, referred to a 'level of secrecy that smacks more of the inner chambers of a Medici princeling than of a public body set up by statute and spending public money' (HC Hansard, 14 January 1998, col. 461). But, this is merely the latest in a series of controversies that

146

have dogged the Corporation in the last year of its life. In June 1997 the High Court ordered TDC to revoke planning permission for an Asda shopping development at Middlesbrough Dock, with Mr Justice Sedley accusing the Corporation of making a 'pervasive departure' from the requirement that planning authorities should objectively evaluate applications. The application was subsequently called in by the Environment Secretary, John Prescott. And at the start of 1998, a regional newspaper highlighted the growing local concern over the issue of 'jobs for the boys' when it reported that TDC Chief Executive, Duncan Hall, planned soon to take up a new post as head of the company managing a Tall Ships Centre in Middlesbrough, a company originally set up by TDC itself to manage one of its own key projects (*The Sunday Sun*, 11 January 1998).

No one who knows the TDC will have been surprised by these allegations and controversies. Over the past eleven years, the TDC, more than any other UDC, has cultivated a style characterised by closed decision-making and excessive secrecy. In so many ways it was a remarkable survivor from the Thatcher years: top-down, single-minded and aggressively pursuing its own agenda. TDC always did things its way – it has certainly not been frightened to 'ruffle feathers' where it deemed it necessary. TDC was little affected by the growing popularity of the concept of partnership in urban regeneration in the 1990s, and has been much more comfortable dealing with the private sector than with local authorities, the voluntary sector or the local community. The TDC had a mission to regenerate Teesside by encouraging the private sector – which always knows best – to get on with it. Critics were best ignored, or told they were simply unable to understand the mission or what Teesside needs. With the TDC, people and agencies were simply categorised: if you weren't 'for' the TDC you must be 'against' it. In many ways, the TDC ethos has not been 'dissimilar to that of the London Docklands Development Corporation in its early days' (Coulson, 1989, p. 10).

This characterisation of the TDC is not exaggerated; we can be confident that it will be recognisable to those who have had dealings with the TDC over the last decade – and to the many who have tried to have dealings with the TDC. In short, the TDC experience illustrates just how powerful and independent a UDC can be. A central argument of this chapter will be that – in the Teesside case at least – the UDC both deliberately excluded local government and other local interests from policy formulation and delivery and also pushed through – with little local modification – a strand of national urban policy heavily based on a 'bricks and mortar' approach. In this sense, we would locate our understanding of the TDC within a traditional 'state-centred' approach, which views quangos as 'executives' of the central state, geared to regaining control at the local level by undermining local democratic institutions (Gurr and King, 1987). We also argue that, in terms of the importance attached to the mediating role of local processes within the 'Locality' debate, TDC's approach has been sustained – rather than modified or contradicted – by 'local sociopolitical milieux' (Imrie and Thomas, 1993b, p. 21).

Few mourned the passing of TDC; it was an organisation which was unloved by most of those who knew it, and little known by the vast majority of people on Teesside. That said, the TDC has delivered some massive projects which have had a major impact on the conurbation. It will, of course, be missed in one way – for its money, which has made possible substantial investment in big projects in unlikely places. It is perhaps easier to develop a judgement about how the TDC has operated rather than about what it has done. Locally, there is some support for what it has achieved, but a review which considers wider impacts and opportunity costs – what might have been done instead – leads to more complex and certainly more critical conclusions. For us, the 'how' and the 'what' are inextricably linked, and will lead us in this chapter to judgements on both the inadequacy of regeneration processes and on the missed opportunities to tackle the real problems of Teesside.

TDC: the local context

Teesside, the conurbation centred on the lower reaches of the River Tees in the North East of England, owes its existence to the industrial revolution. Its principal town, Middlesbrough, was an entirely new town established in the nineteenth century initially to facilitate the export of coal from the Durham coalfield (Briggs, 1968). Iron and steel, shipbuilding and heavy engineering grew rapidly in the latter half of the nineteenth century accompanied by massive urbanisation (North, 1975). The existing small town and river port of Stockton-on-Tees grew and developed, with the built up area of Stockton and Middlesbrough coalescing to form a conurbation with heavy industry at its economic and geographical core. To the north, the nearby town of Hartlepool expanded with the development of West Hartlepool, again a coal exporting port with heavy industry.

In the twentieth century, the process of industrial development continued; ICI established major chemical plants at Billingham and Wilton and further chemical industries were subsequently established on the reclaimed estuarine mud flats at Seal Sands. The combined impetus provided by the Hailsham plan for the North East (Board of Trade, 1963) and Harold Wilson's emphasis on the 'white heat of technology', saw 1960s Teesside earmarked for economic growth, industrial expansion and major investment in infrastructure. The Teesside Survey and Plan aimed to accommodate growth and was infused with the supreme optimism of the time, predicting substantial increases in Teesside's population – from 480,000 in 1969 to 700,000 by 1991 – and the creation of an additional 120,000 jobs (Sadler, 1990, p. 326).

Teesside prospered long after decline had ravaged other places dominated by traditional, heavy industry. Teesside got new investment in steel, engineering and chemicals, heavily supported by regional policy assistance; between 1975 and 1979 Cleveland regularly received over a quarter of the national total of Regional Development Grant payments, reflecting the massive capital investment taking place (Foord et al., 1985, p. 21).

Up until the late 1970s, Teesside was booming. Then came collapse, not just as a result of domestic and international recession but particularly as a consequence of capital investment which displaced thousands of workers, notably in the mainstays of the local economy, ICI and British Steel (Beynon *et al.*, 1994). An area which had been the North East's only big economic success story suddenly went from boom to bust. Between 1978 and 1981, total employment in the Teesside TTWA fell by 13% and unemployment doubled, from 9.1% to 18%. By the early 1980s, Cleveland (embracing the Teesside conurbation and Hartlepool) had become the county with the highest level of unemployment in Britain, with unemployment rates exceeding 20%. A 'quiet revolution' (Foord *et al.*, 1985) had created unemployment and deprivation, leaving a trail of dereliction.

Little was done to help Teesside. The Conservative Government's *laissez-faire* approach meant that industry was left to its own devices, to restructure and survive or go to the wall. ICI shed over 15,000 jobs on Teesside, while the privatised British Steel lost over 20,000 jobs up to the late 1980s. Engineering plants and shipyards closed down as the Government looked ahead to an post-industrial future: manufacturing industry didn't matter. As regional policy was further weakened during the 1980s and the amount of mobile investment diminished, the cash-strapped local authorities in the area, unsure what to do and unable to do much, undertook modest initiatives to encourage small businesses (Hudson, 1990). Training and make-work schemes were set up but had limited impact and were inevitably viewed with scepticism by many unemployed people who questioned their value and purpose: it was tantamount to 'emptying an ocean with a teaspoon' (Cochrane, 1983).

Then, in 1987, the Government announced its commitment to regenerating Teesside by establishing the TDC. This 'second generation' Development Corporation would have by far the largest designated area of all the UDCs, covering almost 19 square miles of land along the River Tees and at Hartlepool docks (see Figure 6.1). A substantial part of the area was derelict, some of it contaminated, although some sites were still occupied by industry. About a fifth of the area was marshland (and has since been set aside as a nature reserve). At designation, there was a very small resident population of around 400 people, mainly at the somewhat isolated settlements of High Clarence and Port Clarence on the north bank of the river Tees near the famous Transporter Bridge.

The Development Corporation's Urban Development Area (UDA) would be created from parts of deindustrialised Teesside hitherto covered by five local authorities (Cleveland County Council and the district councils of Hartlepool, Middlesbrough, Stockton and Langbaurgh). The dominant political culture of local government on Teesside was a moderate and pragmatic labourism, but the new UDC would be located in an area which also had two Conservative MPs in the (then) marginal constituencies of Langbaurgh and Stockton South.

These MPs, Michael Bates and Tim Devlin, were to be keen supporters and lobbyists for the TDC. For example, speaking in 1996, Mr Bates provided a cool welcome for the local authority-inspired Tees Valley Development Company (TVDC) by explicitly comparing it to his 'ideal' regeneration agency, TDC:

> The organisation [the TVDC] is doomed to failure, because it ignores the principles of partnership and co-operation which has made the TDC such a success. (Quoted in *The Northern Echo*, 3 July 1996).

The new quango faced several serious economic difficulties: the key sectors of the local economy were in decline; the region's poor image ensured that it had a low rate of inward investment; while Teesside's particular mix of industries

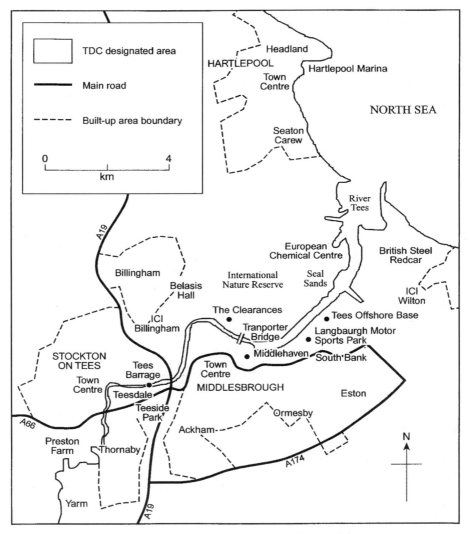

Figure 6.1 *Teesside Development Corporation: the Urban Development Area*

meant that the UDA contained huge areas of derelict and despoiled land.

In November 1987, TDC published an *Initial Development Strategy* which was clearly influenced by such concerns and which served to mark out the direction it was to follow (Robinson *et al.*, 1993, p. 23). The strategy had five main objectives:

- to revitalise the economy of Teesside and create new and lasting job opportunities;
- to attract private sector investment into the area;
- to change the 'image' of Teesside, nationally and overseas;
- to remove dereliction and decay by bringing land and buildings into more productive use while improving the overall environment of the area;
- to improve the quality of life for the people of Teesside.

From the outset, the TDC had a firm belief in the capacity of the private sector and market mechanisms to stimulate the growth and investment that would lead to jobs and improved quality of life in the 'new' Teesside. According to its Chief Executive, Duncan Hall, TDC was to be a 'market-led' and 'opportunistic' organisation (Employment Select Committee, Third Report, 1988, vol. 11, p. 97). That said, it had to be an interventionist institution, allocating substantial public investment to clearing land, supporting private development and changing investor's perceptions about Teesside. As Sadler comments, the creation of a UDC on Teesside reflected:

> the encouragement given to private sector-led intervention by central government as a reaffirmation of the power of the market, counterposed against the restraining hand of local government. (1990, p. 329)

Much was expected of a quango with both money and clout. TDC was also politically well-connected – Mrs Thatcher's favourite UDC was the scene of her famous 'walk in the wilderness' (in Stockton) in September 1987. On that visit, Mrs Thatcher commented 'where you have initiative, talent and ability, the money follows' (*The Journal*, 11 November 1987). Initiative, Talent and Ability became the TDC's abiding slogan and money became the central feature of its logo: 'Tees£side'. Above all, TDC presented itself as an organisation that would get things done, and quickly. As Duncan Hall argued in 1987, ' I am aiming to work myself out of a job and the faster I can do it the better' (*ibid.*).

TDC: the programme

TDC has sought to leave its mark throughout the vast UDA – and carry out its key objectives – by concentrating its development programme on a small number of 'flagship projects' (see Figure 6.1). Many of these schemes were originally identified by consultants Coopers and Lybrand in a report commissioned by the DoE prior to the official TDC start-date of May 1987. Much of the investment and effort has been focused on the major schemes in Stockton

(Teesside Park, Teesdale, the Tees Barrage and Preston Farm) and at Hartlepool Marina. In Middlesbrough, a new football stadium has been built near the river, but plans for the adjacent docklands area have hit major planning problems, so this last major flagship could not be delivered within TDC's lifetime. Further downriver, the Borough of Redcar & Cleveland (formerly Langbaurgh) has received relatively little investment from TDC.

The first major project, Teesside Park, provided an early demonstration of TDC's market-led approach. This large, flat and essentially undeveloped greenfield site, which had formerly been Stockton Racecourse, was the most attractive and readily available site for development. It is highly accessible, alongside the intersection of the A66 and A19, midway between Stockton and Middlesbrough. Despite opposition from the local authorities anxious to protect existing town centres (discussed in the next section), Teesside Park has been developed as an edge-of-town retail centre and leisure complex, which includes a supermarket, shed retail stores, multiplex cinema, night club, restaurants, bingo hall and a health and leisure club. A second phase of retail development, which the TDC wanted to approve, was rejected by the DoE in 1995, partly because of the Government's change of view about retail developments outside existing centres and also since it would have threatened the efforts of Stockton City Challenge to revive Stockton town centre, little more than a mile away from Teesside Park. However, TDC have recently announced that a – reduced – amount of new retail development may still go ahead on the site (TDC, 1998, p. 7).

Teesdale, TDC's biggest flagship scheme and hyped as 'the largest urban regeneration project in Europe, three times as big as Canary Wharf', was the scene of Mrs Thatcher's memorable 'walk in the wilderness' in 1987. Teesdale is across the river from Stockton town centre and was formerly the site of a heavy engineering works. It has been reclaimed and serviced, attracting a mixture of office and housing development as well as a new college of Durham University. There is both private and social housing, together with offices, including Abbey National's mortgage centre and Barclaycard, which relocated from south Stockton. The development of University College Stockton (UCS), which particularly aims to attract local students, was supported by £7.5m from the TDC, while Stockton and Billingham College, the local Further Education College, now wants to relocate alongside UCS. A Research and Development Centre is currently under construction for Kvaerner Process Technology. The Teesdale site remains isolated from the rest of Stockton and TDC's promised 'shopping bridge', modelled on Venice's Ponte Vecchio, which would link Teesdale to Stockton town centre, never materialised. Neither did the Teesside Millennium Building, turned down for Lottery funding, or the enormous ten-storey hotel in the shape of a pyramid, or a proposed theatre. It is, however, hoped that a pedestrian bridge will eventually be provided through an SRB-funded scheme.

Like other UDC's, TDC has been keen on the exploitation of waterfronts for development, but the Tees has required drastic action to turn it into an

Figure 6.2 *Teesdale – new housing and office development next to the River Tees at Stockton*

attraction. At least £50m has been spent on building the Tees Barrage which impounds the river, thereby creating an 11 mile stretch of high water which conceals the unsightly mud, which used to be seen at low tide, and stems the backwash of pollutants from the heavy industry situated downriver. The Teesdale site has thus been provided with a pleasant waterfront, and canal features have been constructed to extend waterfronts into the site. Next to the Barrage itself – winner of a prestigious Concrete Society award for 'creative use of concrete' – a canoe slalom and white water course has been developed, together with a caravan site, and also a pub and restaurant. It now appears highly unlikely that a cable water-ski ('tele-ski') centre at Thornaby, which had been backed by TDC, will go ahead; a local campaign against it and in favour of a park and nature reserve, led the DETR to direct TDC to refuse planning permission in October 1997. The scheme went to a Public Inquiry just before TDC's demise and a decision is now awaited. The TDC's commitment to water-based attractions and activities has resulted in its sponsoring various events including sailing, canoeing, rafting and life-saving competitions on both the Tees and at Hartlepool Marina.

At Preston Farm, TDC has further developed and expanded an existing industrial estate and successfully attracted new economic activity. This 150 acre estate, now almost fully developed, has attracted a large cake factory; the local base for Comcast, which was awarded Teesside's cable franchise; a variety of small and medium-size businesses; a new hotel; and a cluster of car showrooms which TDC calls 'Car City'. A proposal to develop an £11m national shooting club, which TDC supported, came to nothing when the Lottery Sports Council rejected an application for funding – noting that local people were against it.

At Hartlepool, TDC invested heavily in developing Hartlepool Marina and various other developments around the old South Docks. Major infrastructure works were needed for sea defences, dock walls, a yacht lock, roads and piers. The Marina has proved popular and is now believed to be the largest on the north east coast, with 300 berths. Around the Marina, some private housing has been developed and a housing association has built homes on the edge of the site, next to the railway. At a former coal dock, heritage has been packaged into a theme park, 'Hartlepool Historic Quay', a re-created British seaport set in the late eighteenth century, which has proved to be a successful attraction. Nearby, several old ships have been restored and exhibited and a new Museum of Hartlepool has been opened. Negotiations to locate a branch of the Imperial War Museum, focusing on naval warfare, reached an advanced stage but ultimately failed; the £40m project has gone to Trafford Wharfside, Manchester. Asda has moved into the area from the town centre, an American-style factory shopping mall, Jackson's Landing, has been opened and Top Rank has developed a huge bingo hall. A multiplex cinema is also under construction. At the end of March 1988, TDC announced that a local developer, Mandale Properties, would be undertaking further retail and office development at the Marina, would build 100 luxury flats and would also be developing the Tall Ships Centre, previously proposed for Middlesbrough Dock. The 'Historic Quay' and sea defences have been handed over to Hartlepool Council. Like Teesdale, the Hartlepool Marina site is cut off, detached, from the existing urban areas. However, Hartlepool City Challenge has undertaken regeneration of the area between the town centre and the Marina, thus restoring some linkage between the TDC's scheme and the rest of Hartlepool.

Figure 6.3 *Jackson's Landing – factory shopping development at Hartlepool Marina*

The last of the flagship projects is 'The Riverside' or 'Middlehaven', the site incorporating Middlesbrough Dock. Many different schemes have been proposed for this area, including the University College (which, in the event, went to Teesdale), a hospital, a theme park, a theatre, retailing, leisure, offices, hotels, a science park and housing. Middlesbrough Football Club moved to its new Cellnet Riverside Stadium here in 1995 and some infrastructural works have been completed but the future of the rest of the site remains uncertain. TDC wanted to establish a Tall Ships Centre and acquired one sailing ship from the Ukrainian Government. But the development of the Tall Ships Centre appeared to hinge on financing it from the proceeds of retail and warehouse development on the site and then on a bid for Lottery funding which was not successful. As we discuss in the next section, the development scheme put forward by TDC proved contentious, was challenged effectively in the High Court, and the DoE recently ordered that it be subject to a Public Inquiry. A development scheme may eventually go ahead – without the Tall Ships centre, now supposedly destined for Hartlepool – but that depends on securing public and private sector investment.

Elsewhere, TDC has pursued a variety of other projects. In conjunction with the Port Authority, TDC helped redevelop the former Smiths Dock Shipyard as the Tees Offshore Base, a successful centre for offshore-related industry and technology. In Middlesbrough, further development has taken place on Riverside Park, an existing industrial estate which had Enterprise Zone status, and several new businesses have been brought in to the former Haverton Hill Shipyard. New infrastructure has been provided at Teesport to rationalise the port and open up new sites. On the north bank of the Tees, 2,500 acres of land, mostly owned by ICI, has been designated as Teesside International Nature Reserve, which offers wetland habitats and is an important refuge for birds, waders and wildfowl. The newly-created Teesside Environmental Trust, chaired by the former 'chair' of TDC, Sir Ron Norman, has recently been set up to manage and improve the Reserve.

TDC has had little *direct* contact and offered little support to local people and existing local communities. The main exception is housing renewal and associated improvements at The Clarences, where TDC played a key role – and made a real difference. Beyond that, little has been done, apart from small grants to local organisations, and also to individuals involved in sports and the arts. As the spend profile summarised in Figure 6.4 suggests, TDC's regeneration programme has been almost wholly focused on land and property; it has stuck to its original remit.

Notwithstanding the technical and methodological limitations of the output measures used by the DoE to monitor UDC performance (Shaw, 1995), it is clear that TDC has exited having achieved its key lifetime targets (see Table 6.1). From this perspective, much has been achieved:

- Derelict and abandoned land has been reclaimed and brought back into use; by March 1997, 492.2 ha. of land had been reclaimed.

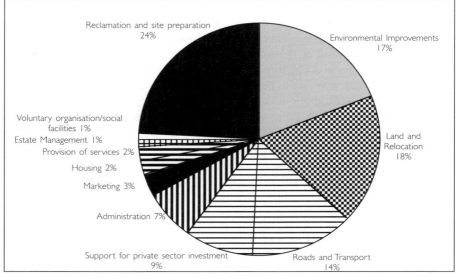

Figure 6.4 *Teesside Development Corporation: expenditure by DoE 1987–1997*
Source: TDC Annual Reports

Table 6.1 Teesside Development Corporation: key output measures

Output	Target 1996/1997	Actual 1996/1997	Total as at 31 March 1997	Lifetime Target
Land Reclaimed (ha)	56.1	57.8	492.2	525.3
Roads (km)	1.6	1.6	27.7	28.7
Housing Completions (no)	167	119	1,306	1,403
Floorspace (sq m)	55,761	69,559	43,1830	470,200
Permanent jobs in new developments (no)	880	2,140	12,226	10,212
Private Sector Investment (£'m)	73.8	74.9	1,003.8	1,090.3

Source: Teesside Development Corporation Annual Report and Financial Statements 1 April 1996 to 31 March 1997.

- New development has been stimulated, with major investment in retail, office, industrial, leisure and housing schemes. TDC claims that over £1 billion private sector investment had been attracted by March 1997.
- In changing the face of Teesside – or, at least, parts of it – and promoting the area, TDC helped create a new, more positive image.
- TDC has contributed to the reconstruction of the local economy and claims to have created 12,226 permanent jobs in new developments over a ten year period.

The physical changes at the flagship sites have been particularly remarkable, even dramatic. The 'wilderness' which was Teesdale has been developed; the

Tees now provides an attractive waterfront; and Hartlepool now has a marina and brings in tourists. TDC has had an impact on the people of Teesside and some of its schemes are both well known and generally popular, notably Teesside Park and Hartlepool Marina. In the last few days of its existence, TDC ran an extensive £60,000 advertising campaign on regional television and radio pointing to its achievements under the slogan 'Leaving a Better Future' and highlighting the major development schemes. According to TDC, 'Teesside has seen a dramatic change for the better'.

But the real value of TDC's achievements is open to question:

- TDC redeveloped fragments of Teesside, creating developments which are physically detached and which do little or nothing for existing communities. In a sense, this was not TDC's fault – the designated area was chosen to provide empty sites, not to incorporate places where people actually are. It has been left to City Challenge initiatives (at Stockton, Middlesbrough and Hartlepool) and SRB programmes to make connections and tackle the problems of existing communities in a more relevant and meaningful way than TDC.
- The approach has been crudely opportunistic; it did not develop from a strategic analysis of what Teesside's needs were but rather what TDC dreamt up and what the market was prepared to deliver. TDC did not believe in planning; indeed, it was ideologically opposed to detailed regeneration master plans, as their inflexibility could lead to a 'rejection of private sector investors' (TDC, 1998, p. 11). The Corporation, like Mr Micawber, much prefers to wait for 'something to turn up'. On the one hand, there has been TDC's fixation with water and, on the other, the market's interest in retail development.
- TDC's job claims may appear, at first sight, impressive. In fact, the cost per claimed job is high (nearly £30,000 per job); some new retail jobs will have led to displacement of existing jobs; and some job-generating developments would have happened anyway.
- The programme has not been integrated into a conurbation-wide strategy. TDC's interest was in filling up sites in its own area, irrespective of impacts on other areas. There is no doubt that retail development, especially at Teesside Park, has had a detrimental impact on existing centres – but that is not something which has concerned TDC.

There is a broader issue: the *opportunity cost* of all this. As Stockton council's leader commented, 'with a budget resembling an international telephone number, I'd expect more' (*Evening Gazette*, 2 April 1998). With over £400m of public money spent during its lifetime (see Figure 6.5), much more could have been done to support the economic and social regeneration of Teesside. If the remit had been different, if the designated area had been different and, specifically, if TDC's philosophy had been different, a much more worthwhile programme could have been pursued.

Such a programme would certainly have paid less attention to the relatively

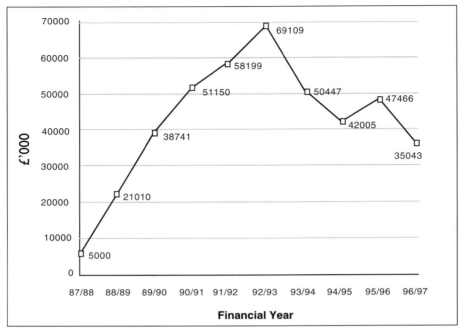

Figure 6.5 *Teesside Development Corporation: annual spending profile 1987–1997*

Source: Teeside Development Corporation Annual Reports

easy option of retail, hotel and sport and leisure developments and attempted the albeit more difficult task of creating more long-term job opportunities for skilled employment in the manufacturing sector. It would also have been much more concerned with developing linkage mechanisms – such as labour market initiatives in the areas of training and skills development – to try and ensure that job opportunities actually 'trickled down' to those in need in deprived communities on Teesside. But such interventions in the local economy were regarded as anathema by TDC and these issues were not addressed. This is made clear in a recent exchange between TDC Chief Executive Duncan Hall and BBC reporter Janet Heaney during the course of a *Money Programme* report on TDC:

> Janet Heaney: How many jobs have gone to people who lost their jobs in the old industries
> Duncan Hall: I can't answer that question
> Janet Heaney: Shouldn't you know whose getting these jobs, isn't that the whole point of regeneration
> Duncan Hall: No, . . . my job is to create job opportunities. It's for employers to determine where they employ the people from.
>
> (*The Money Programme*, BBC 2, 22 February 1998)

If TDC had engaged with local people and acted in partnership with the local authorities and other agencies, a more relevant programme could have been drawn up. Instead, it went its own way. TDC was very much an independent agency which ignored criticism, was insensitive to local needs and which refused to be influenced by local concerns and interests.

The TDC style

Giving evidence to the Employment Select Committee back in April 1988, the TDC's Chief Executive, Duncan Hall reflected on

> the enormous co-operation we have received from local authorities and, frankly, this cannot be overstated . . . there is no coercion involved in the approaches and initiatives of the corporation, sheer matters of co-operation (*sic*). That is totally reflective of the relationship we have got with the county council and the four districts . . . I am not aware of any antagonisms in real terms to the Development Corporation. (Employment Select Committee, Third Report, 1988, vol. 11, p. 101)

There were some signs, at least in the early years, of a pragmatic relationship between some of the local Councils on Teesside and the TDC. This was mainly related to TDC's willingness to undertake extensive marketing campaigns to sell Teesside to potential investors, and particularly its early commitment to develop sites that had already been earmarked for development by the local authorities, but where development hadn't gone ahead, not least because of the increasing restrictions on local government spending. This was true both of TDC's plans for the Hartlepool Marina project – a long-standing Council objective – and the proposed development of the Middlesbrough Dock area. A degree of realism was also found in Stockton Council, where initial resentment concerning the loss of powers to a non-elected quango was partly offset by the fact that the Stockton part of the UDA was to receive a large part of TDC's investment at Teesdale, Teesside Park and Preston Farm.

As the development of the key flagship schemes progressed, the local authority in Hartlepool adopted the most cordial working relationship with TDC. The patron-client relationship (Coulson, 1993, p. 32) between TDC and that Council seems to have been almost entirely based on the former having the resources to allow the latter to realise its long-standing ambition – first mooted in the 1970s – to develop a marina in the town. Both organisations are proud of the marina 'flagship', recently described by the Development Corporation as 'an accepted phenomenon' (TDC, 1997, p. 6) and, by a local newspaper, as the 'jewel in the crown' of Hartlepool's renaissance (*The Northern Echo*, 9 July 1997). Links between the two organisations, at a senior level, were also important, and the Labour leader of Hartlepool Council, Bryan Hanson, has been a long-serving member of the TDC board. As TDC approached exit in March 1998, only the Hartlepool leader publicly praised the Corporation in the region's media, while the leaders of both Stockton and Redcar and Cleveland Councils condemned it.

Yet even in Hartlepool, the relationship with TDC could hardly be described as close. According to one officer, discussions with the Corporation were mainly on 'points of detail rather than substance', and the relationship was 'very much on TDC's terms'. The genuine Council support for TDC's role in raising the image of Hartlepool over the last decade needs to be placed alongside the more critical view – expressed by a council officer – that in doing so, the TDC was 'more concerned to market itself than the town of Hartlepool'.

The considerable deterioration in TDC's relationships with the other local authorities during the 1990s seems to suggest that the Chief Executive's comments to the Employment Select Committee are perhaps best viewed as the over-optimistic judgements typical of an organisation in its 'honeymoon period'. Indeed, the recent views of a Teesside MP in the House of Commons seem a more appropriate summing up of the last ten years of TDC-local authority relationships:

> It is a well known fact of life on Teesside that the TDC and particularly the corporation's chief executive, Duncan Hall, are difficult to deal with. It is also a well-known fact that many Teesside local authorities, both past and present, have found it difficult to work constructively with the TDC. (Ashok Kumar MP, HC Hansard, 14 January 1998 Col. 460)

Indeed, one theme that emerges from the Teesside case is that relations between the TDC and the local authorities got worse over time, rather than better. It seems that the more local councillors and officers got to know of the TDC, and how it operated, the less they liked what they saw. In the following case studies, we highlight three particular areas where the local authorities on Teesside have found it hard to work with the Development Corporation.

Case study 1

In Langbaurgh (now Redcar and Cleveland), conflict occurred over a proposed Motor Sports Complex on an old blast furnace site in the South Bank area. The Council's original plan was to develop a motor sports park for low-cost karting and autocross which could also host commercial events such as Formula 3 racing. The scheme was adopted by TDC in 1988 and the first stage development involved the Corporation developing the go-kart track. However, a crucial change of emphasis followed from TDC's decision to appoint private companies to develop and manage the project. Hence, the eventual plans for the Formula 3 circuit were now to be combined with a wider range of developments including a hotel, retailing, bowling and roller skating and a bingo hall. This is early evidence of what was to become a TDC characteristic – schemes emerged from the 'ether', then disappeared just as suddenly, only to be superseded by new, often radically altered, proposals. In this case, the new plans were strongly opposed by Langbaurgh council on environmental grounds, because of the expected noise nuisance to local residents and this opposition spilled-over into the local council elections.

Opposition to the scheme also came from Cleveland County Council. In the end, the developer withdrew the scheme, much to the annoyance of the TDC. The relationship between Langbaurgh and TDC appears never to have recovered from this acrimonious dispute and the Borough has subsequently had little TDC investment.

Case study 2

Despite being the recipient of several key schemes, Stockton Council's relationship with TDC has become increasingly uneasy – the subject of suspicion on both sides. One of several issues generating difficulties has been the conflict and disagreements over the development of Teesside Park. Initially, the Coopers and Lybrand consultants' report had suggested that the former Stockton racecourse site was suitable for leisure developments (including an all-weather racing track), a hotel and a modest themed retail element. However, once TDC had accepted a private developer's plans, which included translating the retail element into a large off-centre 'retail warehouse' scheme, strong opposition emerged from Stockton Council, concerned about the displacement effects on Stockton town centre. As the Council pointed out in 1992 – after companies like Poundstretcher and Texas had closed down their town centre outlets and moved to the Park – 'most of the retailers are already represented in local town centres and also compete directly with local town centre outlets' (Stockton MBC, 1997, p. 7). Disagreement continued as Stockton formally objected to TDC's proposed 100,000 sq. ft. extension to the retail park; the proposed extension was eventually rejected by the Department of the Environment and the revised application was held up by the Highways Agency because of concerns with site access. It is ironic that while TDC was seeking to expand retailing at Teesside Park, Stockton's City Challenge initiative has had to focus its efforts on regenerating the ailing Stockton town centre, already hit by competition from Teesside Park.

Case study 3

As noted in the introduction to this chapter, there has been considerable conflict recently over the development of the 150 acre Middlehaven/Riverside site in the Middlesbrough Dock area. After several false dawns – most notably the University College project that was eventually developed in Teesdale – and Middlesbrough Council becoming increasingly concerned at the lack of progress, TDC announced proposals in early 1996 for an Asda hypermarket, retail warehousing, restaurants, offices, a hotel and a Tall Ships Centre. Despite Middlesbrough's desire to see the long-awaited development of the site take place, the Council's eventual concern over the effects of the retail developments on existing shopping outlets in the town centre was sufficient to delay the scheme, and was strongly supported by Morrison's, whose supermarket development at Berwick Hills would be directly threatened by the Asda store on Middlehaven. Redcar and Cleveland Council was also very critical, being concerned about the impact on its own shopping centres such as South Bank. Despite TDC's recourse to brinkmanship – the Corporation argued that without the finance raised by the retail developments the project could not go ahead, and the proposed Tall Ships Centre would have to be transferred to Hartlepool – the protests of WM Morrison and Redcar and Cleveland were upheld in the High Court in May 1997 in a decision that was very critical of both TDC's judgement and its disregard of the views of other local agencies. After some early signs that the TDC board were keen merely to re-approve the proposed Asda development, continued pressure from Redcar and Cleveland Council saw the application called in by the new Secretary of State, John Prescott. This, combined with Middlesbrough Council's increasing concern to get some development on Middlehaven, has resulted in a compromise proposal emerging which removes the hypermarket development and restricts the retail warehouse element to non-food outlets. This received the guarded approval of the interested parties in December 1997. However, by January 1998 controversy had returned to the Middlehaven scheme, when the local media, two local MPs and Redcar and Cleveland Council called for an inquiry into allegations that TDC's chief executive had been appointed to run the proposed Tall Ships Centre (*The Sunday Sun*, 18 January 1998). With just two days to go before closing down, TDC announced that the Tall Ships Centre would, instead, be developed at Hartlepool Marina as part of a scheme put forward by a Teesside development company, Mandale properties. This has left Middlesbrough Council in the lurch, desperately trying to find backers to redevelop Middlehaven (*Evening Gazette*, 31 March 1998).

These three cases capture the essentials of TDC's style; the approach it brought to its dealings with other local agencies on Teesside. The main features of this approach have been:

- secrecy and a lack of openness;
- unforeseen changes in plans;
- a desire to hurry proposals through without adequate consultation;
- brinkmanship, often involving playing one authority off against another;
- the singular pursuit of its own interests – development always happened on TDC's terms;
- a general disdain towards local authorities or organisations involving local government interests.

This final point is worth reinforcing. TDCs relations with the other agencies of local governance have also been problematic. Thus, the TDC had a poor relationship with the Tees Valley Joint Strategy Unit which was set up on the advice of DoE Circular 4/96 to carry out strategic planning after the abolition of Cleveland County Council. The Unit's views on the Middlehaven scheme were disregarded by TDC on the grounds that they had 'no formal status', a view heavily criticised by Mr Justice Sedley in the subsequent High Court ruling (Baber, 1997, p. 10), while the Unit's recent requests to be consulted over the preparation of TDC's Exit Strategy have been ignored. Indeed, such was the general lack of consultation with local agencies over TDC's exit strategy that Environment Minister Angela Eagle was forced to admit in the House of Commons that:

> I am surprised that, unlike other Development Corporations, TDC has decided not to enter into any agreements with the local authorities on Teesside for them to take on packages of assets and liabilities . . . I would have liked to have seen wider discussions than those that have taken place. (HC Hansard, 14 January 1998, col. 465)

In TDC's final days, the local authorities and other agencies struggled to find out what would be happening with TDC sites after exit. Given the lack of information offered by TDC itself, the local authorities had little more to go on than rumours, and were left awaiting the details that would emerge after TDC's demise. It is known that the Commission for New Towns is taking over the Tees Barrage but it is not known which underdeveloped sites have been sold off, or on what terms.

A revealing minor turf war recently developed with the Tees Valley Development Company (TVDC), set up by the local authorities in 1996 to attract investment to the area. Both the Development Company and TDC were thanked by the Abbey National Bank in the formal announcement of their plans to build a new mortgage centre in Teesdale, creating some 200 jobs. TDC claiming the bulk of the credit, quickly issued a press release criticising TVDCs 'unwarranted and misleading claims of involvement' (*The Northern Echo*, 16 January 1997).

TDC's relations with the local communities in and around the UDA have never been close. Indeed, the Corporation has generally been viewed by the community as being distant, detached and unapproachable; not surprising, since TDC spent little time or money on ensuring any community involvement in the process of regeneration. By 1998 very little had changed from 1993, when we noted how, in its first five years, TDC had appeared

> to regard community participation as irrelevant or even meaningless. Since developers will know their market and make the decisions there is little justification for community participation. Thus the Chairman [Ron Norman] said that he feels that 'consultation leads to muddle and delay'. (Robinson *et al.*, 1993, p. 51)

The general isolation of the Development Corporation from the community – 'people recognise the letters 'TDC' but it doesn't touch the lives of people in this community' (voluntary sector worker quoted in Robinson *et al.*, 1993, p. 51) – did not prevent conflicts arising when community groups were confronted with the TDC's style of operating. As recently as 1997, TDC incurred the wrath of community groups on three particular occasions:

- TDC pushed ahead with plans to support an £11m National Shooting Centre at Preston Farm, despite widespread opposition from local people in the wake of the tragedy at Dunblane. The scheme was eventually thrown out in February 1997, when the National Lottery Sports Council, who had been asked for a grant of £7m, noted that, 'it isn't even in the running for £7 let alone £7m . . . why on earth would we support a scheme that was going to be so unpopular local people would not use it?' (*The Northern Echo*, 4 February 1997).

- In June 1997 angry residents who live on the Teesdale site in Stockton formed a residents' association to fight TDC's plans to build five-storey office blocks which would restrict their view of the pleasant canal feature at Teesdale. According to the residents' spokesperson, the Corporation had assured them, when they moved in, that any offices built near their homes would be only three floors high. Moreover, they had not been consulted on the present scheme, put forward by Mandale Properties, and had been given only two weeks to object (*The Northern Echo*, 5 June 1997).

- In October 1997, a wide range of community groups, including local environmentalists, raised a 5,000-name petition against TDC's proposed £1m overhead cable water-ski (tele-ski) scheme, arguing that it would destroy wildlife in one of the few areas of unspoilt countryside in the Stockton-Thornaby area. After receiving the petition, the Environment Secretary John Prescott ordered a public inquiry. On the first day of the inquiry, in March 1998, TDC was accused of misleading the public when it was discovered that the plans displayed involved a far larger ski centre than was first suggested. Thus, the chair of Thornaby Town Council, was 'amazed to see on the latest plans extra car parks, two ski runs including a ski ramp for jumps, more buildings and a dry ski training run' (*The Northern Echo*, 4 March 1998).

In the next section we will attempt to highlight the key factors that help explain TDC's failure to mesh with other local institutions and interests, and how its particular approach to local governance on Teesside was influenced by the combination of general and – more distinctive – local factors.

The TDC and patterns of urban governance

> The overall pattern of relations between local authorities and local quangos is one of variety and complexity. In practice, relationships between each species of quango and local authorities can fall anywhere on a continuum between co-existence and conflict, depending on the specific configuration of political, social, economic, historical and geographic conditions. (Greer and Hoggett, 1996, p. 165)

Given the arguments contained in the previous section, we would obviously locate the overall pattern of TDC-local authority relationships firmly towards the 'conflict' end of Greer and Hoggett's continuum. Despite the welcome development of more sophisticated approaches to understanding UDCs and local governance, on Teesside, at least, there is little evidence of the TDC 'going native', 'meshing with local interests', 'modifying original policy objectives' or, gradually, becoming 'embedded' into the locality. Indeed, the Teesside experience suggests that this UDC consistently served as an arm of a national (unreconstituted Thatcherite) urban policy, implemented a predominantly private sector-led bricks and mortar approach and acted to exclude local government and other local interests from involvement in the process of regeneration. TDC's approach was not affected by the passage of time – indeed, as we have argued, relations with local agencies actually worsened during the 1990s – suggesting perhaps the need to invent a new term, 'un-embeddedness', to describe this outcome.

What is the explanation for this – arguably distinctive – pattern of local governance on Teesside? Put simply, why did TDC act as it did and how was it able to effectively ride 'roughshod' over a wide variety of local interests on Teesside? Our argument emphasises the particular impact of a powerful quango, dominated by a Chief Executive and Chair with particular views about 'regeneration', in an area traditionally accustomed to the exercise of corporate power and characterised by an often weak and fragmented local political system.

TDC wasn't just a powerful quango with considerable development powers. It was a powerful quango dominated by two key individuals, its Chair, Sir Ron Norman, and especially its Chief Executive, Duncan Hall. The former served as the public face of the Corporation while the latter was the less public strategist, providing leadership within the TDC and largely determining its approach. As Andrew Coulson noted (a decade ago) it is predominantly these individuals that defined the TDC style:

> the chairman has a property background and has interpreted his role as getting physical development, bricks and mortar, on the ground, as quickly as possible. Its Chief Executive was previously in charge of Corby District Council, has little faith in traditional town planning, gives little weight to consultation and involvement of local people, and believes that in most traditional local authority economic development activity 'the fundamental requirements of economic rejuvenation, namely political certainty and decisive decision-making, to match the requirements of incoming industry, are irrevocably lost'. (Coulson, 1989, pp. 9–10)

The combination of two assertive personalities, both committed to utilising private property interests in the rapid transformation of industrial Teesside – and with little time for traditional 'ways of working' – ensured that TDC appeared to many as an autocratic and top-down organisation. According to the leader of Stockton Council, the TDC's Chief Executive was 'running a one man band . . . even his own people don't know what's happening 97% of the time' (Councillor Bob Gibson quoted in *The Evening Gazette*, 11 March 1992). The top-down way that the Development Corporation operates also, necessarily, seems to have limited the input and influence of the TDC board which is, in any case, further weakened by the lack of local knowledge of several of its members.

This powerful quango – with its distinctive approach to urban development and urban governance – was superimposed on a particular local socio-political structure on Teesside. However, in the case of TDC, such local factors tended to reinforce (rather than redirect) the thrust of non-local pressures, and the TDC's autocratic style and private sector-led approach found less effective and less co-ordinated local opposition on Teesside than one might perhaps expect. Two relevant factors can be highlighted.

First, it may well be that Teesside's cultural experience of large and powerful paternalistic employers – particularly exemplified by the corporate power of ICI (Turner, 1968) – has much to do with both the sanctioning of a predominantly private sector-led approach and the lack of political and popular pressures for wider participation. The corporatist, top-down style of decision-making that characterised the post-war consensus between private capital and the local state on Teesside became firmly established (Sadler, 1990). Teesside's historical legacy has meant that TDC was able to operate very much like a large and dominant company, pushing forward schemes without explanation, not facing any serious challenges from local people and able to pull out (exit) on its terms, without recourse to other local interests.

Secondly, TDC was imposed on an already fragmented local government system, with power shared – often very uneasily – between Cleveland County Council and the four District Councils. The different tiers of local government often failed to co-operate effectively in areas – such as economic development and planning – where there were overlapping interests. There were also genuine conflicts and rivalries between the Districts themselves, which both prevented the emergence of any coherent economic strategy supported by all the local authorities in the area, and made a co-ordinated response to TDC and its schemes difficult to organise. In the absence of a clear local economic strategy – and faced with often muted political challenges from individual councils – TDC was able both to claim that it was the only organisation with a clear vision of the 'New' Teesside, and to employ an often very successful strategy of 'divide and rule' towards the different local authorities.

Conclusion

It is not difficult to see why the TDC is said to have been Mrs Thatcher's favourite UDC, or why, in the 1997 General Election campaign, she returned to Teesside to view its achievements – 'Good Conservative policies translated into practice'. It was the kind of organisation that came up with solutions rather than problems: TDC got things done and didn't waste time on consultation or on detailed analysis of what was needed. And, in Teesside, the institutions of local governance (and also local community institutions) were too weak, or perhaps lacked the understanding or confidence, to effectively challenge the TDC.

TDC will be remembered as much for the *way* it operated as for what it actually *achieved*: it has left its particular imprint on the local political culture of Teesside. TDC has not left behind an empowered local community in which local people are able to harness their knowledge and experience in developing a sustainable approach to urban regeneration. If anything, a decade of the

TDC style of governing has actually served to undermine the capacity of local agencies and local communities on Teesside to develop the very networks and partnerships that are now viewed as integral to successful regeneration practice (Lowndes *et al.*, 1997).

There is no doubt that the institutions and the local people of Teesside are pleased that the money was spent there, not elsewhere; that the river Tees is now much more attractive; that Teesdale is no longer an eyesore; and that Hartlepool Marina has become a tourist attraction. But it is a curious form of regeneration. The money has been spent, not where people are but, instead, on starting again – building a new Teesside and ignoring what exists already. The money has been spent, not on regenerating communities blighted by high unemployment and deprivation, but on creating new, detached, developments in unlikely places. TDC pursued its remit, did what it was asked to do. It both illustrates the particular limitations of the UDC approach, and provides a stark demonstration of what was wrong with the wider Thatcherite agenda for 'those inner cities'.

Further reading

A good beginning, for anyone interested in understanding the development of industrial Teesside, is the chapter on Middlesbrough in Asa Brigg's *Victorian Cities* (1968). The contributions by Sadler (1990) and Hudson (1990) also provide additional historical perspectives on economic development, political modernisation and corporate power. Within the recent literature on Locality studies, Beynon *et al.*'s, *A Place Called Teesside* (1994), provides a very comprehensive account of the impact of global restructuring. The introduction of the UDC on Teesside is discussed by Andrew Coulson in his INLOGOV report of 1989. Our *More than Bricks and Mortar* report (1993) provides a comparative assessment of both TDC and Tyne and Wear Development Corporation after their first five years. A very detailed picture of TDC activities in the Stockton area is provided in the annual Progress Reports produced by Stockton Council. Recently, TDC has produced a Regeneration Statement (TDC, 1998), which provides both a useful insight into the Corporation's overall philosophy and a short summary of its key schemes. The CD-ROM version of the main regional daily newspaper covering the Teesside area, *The Northern Echo*, is also a good source of material on the TDC and its activities. In particular, the paper has covered many of the recent controversies over TDC's plans for Middlesbrough Dock and its Exit Strategy during the period 1996–1998.

7

Urban Policy in Sheffield: regeneration, partnerships and people

GORDON DABINETT AND PETER RAMSDEN

The early 1980s recession tore through the fabric of the Lower Don Valley like a whirlwind, destroying jobs, factories and Sheffield's sense of pride in the achievements and traditions of its industrial heartland. (Martin Liddament, previously Corporate Affairs Manager for the Sheffield Development Corporation, in Hey *et al.*, 1997, p. 99)

The successful regeneration of the area has brought with it a new spirit of partnership in the City and a new confidence in the future. Both the Valley and the City are now much stronger and more able to attract the employment that is so necessary to generate wealth required to improve the quality of life of its citizens. (Sir Hugh Sykes, previously Chair of the Sheffield Development Corporation, in Hey *et al.*, 1997, p. 143)

Introduction

During the relaunch of urban policy in 1988, urban development corporations were described as the most important attack ever made on urban decay (DoE, 1988). Such a bold political claim throws open the debate about the role of UDCs to include fundamental questions about the nature of urban policy. In particular, urban policy since its early conception in the 1960s has addressed issues of inequality between urban and non-urban areas, but, more significantly, within cities.

This chapter describes the delivery of urban policy in one specific locality over the decade 1986 to 1996, and in particular assesses the legacy of imposing a UDC in 1988. Sheffield was declared an urban programme authority in 1979–80, and in those early days it also took an antagonistic stance towards central government policy on urban local authorities and developed alternative strategies and approaches. Later, the city was to fail in two bids for City Challenge funds in 1991 and 1992, but was then three times successful in obtaining Single Regeneration Budget (SRB) challenge funds in 1995, 1996 and 1997. Against this background, the City Council, which had developed clear local policy responses to urban decline during the early 1980s, had by

168

1996 become entrapped by financial debts and firmly committed to 'partner-ship working'. Over the decade examined here, total employment in the local authority declined from about 30,000 to some 20,000 employees; the city hosted the World Student Games (WSG) in 1991; saw the construction of a £240 million light rail system, the South Yorkshire Supertram; and the evolu-tion of a new form of urban leadership, the Sheffield City Liaison Group (SCLG). Thus the nature and scope of urban policy, urban regeneration and urban governance all become entangled in this story of Sheffield. By examin-ing the operation of the UDC within this overall experience, we claim that urban policy must always keep sight of its more basic objectives – to assist people and communities in improving their quality of life by providing appro-priate locality based assistance.

Urban policy – the Sheffield context

To understand the significance of the UDC declaration in Sheffield requires an appreciation of the historical context of urban policy in the city and in partic-ular the Lower Don Valley, the area covered by the UDC (see Figure 7.1). Sheffield is the major urban centre of the South Yorkshire sub-region. This region does not form a cohesive conurbation in geographical terms, but was united by the traditional industries of coal, steel and engineering. During the late 1970s Sheffield was regarded as an area of relative prosperity in this sub-region. Sheffield, like all major urban areas, has always had a number of res-idents who suffer social, economic and political inequality. In relative terms, these groups were small and concentrated in a number of well defined spatial areas, and urban programme funds were used after 1979 to tackle this urban-based inequality.

It can be argued that Sheffield only became strongly associated with depres-sion and urban decline after 1981 (Dabinett, 1995). At this time confidence within the city was very low. There was no new investment, housing and land markets were stagnant, employment falling and unemployment rising, and the main area of traditional employment, the Lower Don Valley, was becoming an enormous wasteland of vacant buildings and derelict sites (Hey *et al.*, 1997). The level of unemployment in the city rose above the national average after a long period of stability. This was a direct result of job losses in steel, steel pro-cessing, engineering and cutlery (Tweedale, 1995). Therefore, the city's posi-tion changed rapidly and relatively recently, and this was explicitly linked to the state of the local economy.

This situation was fuelled by the conflict between central government and the City Council about local government funding and local taxation. With David Blunkett as leader of the council, Sheffield was at the forefront of the national campaigns to save local authority jobs and services (Blunkett and Jackson, 1987). In the 1980s the local authority refused to become involved in the deregulation experiments of Enterprise Zones and early declaration of UDCs, and did not receive special urban policy status through Task Forces or

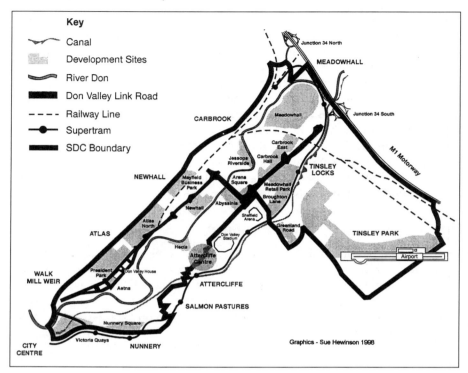

Figure 7.1 *The Lower Don Valley – Sheffield*

City Action Teams. Conflict and division grew between the controlling Labour nexus (the council, the party and the wider movement) and private capital interests in the city (Seyd, 1990). The Lower Don Valley itself became a powerful political symbol of this debate, as the issues incorporated the nature of economic policy as well as public expenditure controls.

This local economic and political situation was to change again late in the 1980s coinciding with the election of a further Conservative government, which saw the period 1986 to 1991 become distinguished by economic regeneration and partnership rather than industrial bargaining and old style corporatism (Lawless, 1990). It was with the declaration of a UDC in 1988 that the main instruments of national urban policy were to be implemented in the city. Up until then, Sheffield remained an urban area where local policy responses dominated and central Conservative government urban initiatives were marginal, unlike their policies towards public expenditure and industry. The relationship between central and local policy was to take another turn in the mid-1990s when the city secured some £91million of SRB Challenge funds for the period 1995 to 2004. The main policy measures taken in the city are described next, concentrating on the period 1986 to 1991 since this coincides with the declaration of the UDC and changes in local policy.

Sheffield was first designated an urban programme authority in 1979 and it consistently funded a breadth of projects generally of a small-scale nature yet orientated to specific local needs. Such approaches have been shown to be relatively effective in delivering local benefits and job opportunities (Turok and Wannop, 1990). Table 7.1 shows the basic breakdown for the bids made by Sheffield City Council for 1986–89 and 1991–94. The 1986–89 programme was the first to recognise explicitly the economic restructuring that had occurred in the city (Sheffield City Council, 1986). This programme included a supplementary bid for activities in the Lower Don Valley, and other major projects included proposals for a science park, a technology park, as well as direct grants to firms, either in Industrial Improvement Areas or for small businesses. Despite this reorientation of the bid in 1986, 32% of expenditure was still allocated to support social initiatives, and only 34% was to be spent on economic development capital projects (see Table 7.1).

This was in contrast to the much smaller bid made five years later (see Table 7.1), in which expenditure for social initiatives accounted for only 22% of the total bid, and projects seeking economic objectives made up 52% of the bid (Sheffield City Council, 1991a). With the exclusion of the Lower Don Valley, the 1991–94 bid focused on the wider economic strategy of the City Council, and in particular the attempts of this strategy to 'spread the benefits' of regeneration to the community as a whole. Some of the major projects in this bid included an Afro-Caribbean enterprise centre and an employment project on the Manor, a local authority housing estate. However, other major events were to have an impact on urban regeneration policy in the city between 1986 and 1991 including the staging of the WSG (Dabinett and Lawless, 1998).

The WSG has been the most controversial of the initiatives in Sheffield. Very little research was done into the probable costs and benefits that the WSG would bring to the city at the time the decision was made (Foley, 1991). The construction of new facilities included an arena and a stadium in the Lower Don Valley and two new swimming pools. A cultural festival was also held and the Lyceum Theatre was renovated at a cost of £12 million. Overall,

Table 7.1: Sheffield Urban Programme bids 1986–9 and 1991–4.

Main objective	Total Expenditure 1986-89 (£000)	Share of expenditure 1986-89 (%)	Total Expenditure 1991–94 (£000)	Share of expenditure 1991–94 (%)
Economic	4,122	34	3,947	52
Environment	2,931	24	1,478	19
Social	3,927	32	1,721	22
Housing	1,167	9	455	6
ALL	12,147	100	7,601	100

Source: Sheffield City Council (1986; 1991a)

the cost of the construction work has been estimated at £147 million with relatively few financial benefits or contracts for Sheffield and region (Sheffield City Council, 1990a). The management of the WSG ran into severe problems when the public/private company set up to run the event ceased trading in June 1990. The Council subsequently took over the direct management of the WSG at an unknown cost, although estimates suggest the total cost to the City Council had risen to £20 million per annum by 1995 (Dobson and Gratton, 1996). The WSG went ahead in 1991 with some commentators claiming great benefits to Sheffield in terms of morale, image, new facilities and the promotion of Sheffield as a 'city of sport'. Opponents pointed out the cost to the local taxpayer, the concurrent closure of schools, local sports facilities and libraries, and the élitist nature of the new facilities.

Since the development of an economic strategy by the City Council set the context for many policy matters during the period 1986 to 1991, it is worthwhile briefly considering the nature of this strategy. The concept of an economic strategy began to emerge in 1981, with the formation of an Employment Department and Employment Committee (Benington, 1986). This was a clear political response to deepening industrial decline and was essentially employment based rather than a coherent industrial or regeneration strategy (Totterdill, 1989). However, in 1986 the Council was forced to concede that private sector investment would be necessary to meet the shortfall between available public resources and the investment needed to complete agreed projects. It was believed that a new development strategy and agency were required in order to secure such investment. The agency was seen to be a public/private partnership which could change the anti-business image of the Council, representing a significant move away from municipal socialism (Lawless, 1990). This turned out to be the Sheffield Economic Regeneration Committee (SERC) with substantial private and voluntary sector representation. SERC was charged with overseeing the development of a collaborative approach to regeneration in the city, at that point of time outlined in the Twin Valleys Strategy. This strategy was intimately linked to Urban Programme and European Commission funding, the latter being an ever increasing influence on local regeneration policy in the city during the late 1980s and 1990s.

Later, the City Council began to recast this strategy to incorporate other geographical areas of the city and to broaden the issues tackled. The outcome was Sheffield 2000, a City Council-led initiative to develop positive proposals for the city through to the year 2000 (Sheffield City Council, 1990b). The organisation of this strategy was also partnership based and represented an extension of SERC's operations. The initiative may be regarded as an attempt to reclaim the policy areas ignored by the UDC and Conservative urban policy, but the approach was managerial in its style and operation, and Sheffield 2000 represented more of a corporate framework rather than an alternative employment, economic or social policy or plan. It lacked resources, a city-wide consensus and broad support in the local authority.

The need to search for new, and additional, funding to implement Sheffield

2000 was well illustrated by the unsuccessful City Challenge bids prepared in 1991 and 1992, both containing proposals extracted from this strategy (e.g. Sheffield City Council, 1991b). Spatial targeting was again a feature of these bids which were largely based on site-specific projects. The first bid had a combined capital and revenue cost of £180 million, of which the private sector was to provide 74%. This bid in many ways illustrated the weaknesses of the strategic approach adopted in the city. It was criticised for being over-ambitious, in particular with respect to the level of private sector funding for which there were few precedents in the city. It was also argued that the bid had been prepared without full or proper consultation, relying on the new corporatism of the city's organisations such as SERC for the formulation and implementation of proposals, rather than using or setting up grass-roots participation.

Thus, 1992 came to represent another watershed in the evolution of urban policy in the city. Whilst SERC had widespread representation from within the city, by the mid-1990s its role had been reduced to one of simply briefing these partners on broad issues. As a result, a new partnership grouping emerged in 1992, the Sheffield City Liaison Group (SCLG). This took a more executive remit and initially had a narrower membership drawn from five constituencies: the local authority; higher education; the private sector as represented by the Chamber of Commerce and the Cutlers Company; the Sheffield Health Authority; and the local development agencies – the Sheffield UDC and Sheffield TEC. Significantly the Group claimed that it did not represent every interest in Sheffield. 'We have deliberately remained a small group, with a focus on action, in order to avoid the danger of becoming a talking shop . . . In short, the role of the City Liaison Group is to provide leadership' (Sheffield City Liaison Group, 1995, p. 3). The successful attraction of SRB Challenge Funds might be seen as the first outcome of this shift in the City's partnership arrangements (Dabinett and Lawless, 1998). Clearly then, the operating lifespan of the SDC encompassed many fluctuations in local urban policy, and the evolution of new relationships between the City Council, central government and local business interests.

The Sheffield Development Corporation

In June 1987 SERC commissioned an independent report on economic regeneration in the Lower Don Valley. The City Council hoped that the report would convince the Government that a considerable amount of public money should be invested in the area and that the local authority could control the process, along similar lines to the Heartlands in the West Midlands. The report was published in November 1987 and recommended that an Urban Regeneration Authority should be established to oversee the regeneration of the valley (Coopers and Lybrand, 1987). On 7 March 1988 the Department of the Environment (DoE) announced the proposed establishment of a UDC to be located in the Lower Don Valley as part of the third round of designations. The order establishing the Sheffield Development Corporation (SDC) was approved by Parliament on 29 June 1988. The corporation was intended to have a life of seven years with £7 million to spend each year. In the event, the Central Government grant and the timescale were both increased. The SDC was wound up in 1997, having received over £100 million in grant aid. The City Council opposed the setting up of a UDC but decided to negotiate terms rather than take their opposition to the House of Lords as Bristol were to do. The result was an agreement over consultation, a commitment to Section 71 of the Race Relations Act and a commitment to discuss possible agency agreements with the City Council. In addition three councillors took positions on the SDC board. A Community Director was also appointed, being the first such post (Kirkham, 1990).

The SDC was directed to adopt the comprehensive and integrated approach to regenerating the Lower Don Valley recommended by the consultants the previous November, and its overall objective was to service the economic and physical regeneration of the Lower Don Valley at minimum cost to the public sector, levering maximum private investment into the area. The area covered by the SDC was 2,000 acres of land (see Figure 7.1), of which 35% was derelict or vacant at declaration. It included some 800 firms employing 18,500 people, predominantly in metal manufacturing and engineering. There were only about 300 residents in the area although the Valley was surrounded by areas of acute poverty and high unemployment (Kirkham, 1990).

The declaration of the SDC represented a significant development in the city, in that it acknowledged the scale of the problems in the Lower Don Valley, the perceived requirement for new organisational structures, and the need for additional funding which would not have been available to the city without the declaration of a UDC. The SDC formulated its plans for the Valley through its own corporate planning structure and through the publication of Planning Frameworks (SDC, 1989). These documents promoted the property-led and the flagship approaches to regeneration, but it must be remembered that many of the major schemes, including the Meadowhall shopping centre, proposals for an airport and the Canal Basin office and leisure scheme, were initiated before the SDC was established.

In the early years, the majority of expenditure went on the purchase of old steel works sites in the Valley. Table 7.2 shows that the largest single expenditure was on land acquisition (some 35%), with smaller sums spent on transport infrastructure, site preparation and grants to business. Initially the SDC took the view that private developers would be willing to undertake all site preparation, reclamation and servicing works. Thus, the main activity in the first years focused on acquiring land and the building of a major spine road to provide access to the new sites. This decision led the SDC to declaring some 200 Compulsory Purchase Orders (CPOs) on local businesses in 1989. This was to lead to a lot of acrimony between the Development Corporation and local companies, which only receded when the road proposal was reduced in scale and most of the CPOs lifted (Raco, 1997). The multiple ownership of the

Table 7.2: Sheffield Development Corporation spending 1988–1996

£m	1988–90	1990–1	1991–2	1992–3	1993–4	1994–5	1995–6	Total at March 1996
Land acquisition	12.5	11.8	3.8	5.3	1.3	2.0	0.6	37.3
Site preparation	0	0.9	3.6	1.2	4.1	2.0	0.9	12.7
Transport	0.4	1.9	0.4	2.4	5.1	3.9	3.7	17.8
Environment	0.4	0.4	1.2	0.6	0.8	1.4	0.8	5.6
Business grants	0.3	1.3	1.5	0.7	1.7	3.3	2.3	11.1
Community	0.1	0.1	0.2	0.2	0.1	0.1	0.1	0.9
Promotion	0.5	0.4	0.4	0.4	0.4	0.4	0.2	2.7
Administration	3.2	2.2	2.2	2.4	2.5	3.0	3.0	18.5
TOTAL	17.4	19.0	13.3	13.2	16.0	16.1	11.6	106.6

Source: SDC Annual Reports 1989/90 to 1995/6.
NB – SDC began operation in July 1988, figures for first year cover period from then until March 1990.

proposed development sites in the Valley did pose a barrier to effective regeneration, and land acquisition was often necessary to bring sufficiently large sites forward for development. This approach, in part, reflected the market philsophy of central government and the SDC, but also the buoyant state of the market in 1988/1989. Two years later and the SDC had to review this strategy as the severity of the recession became evident and property markets remained flat. Thus in 1991/1992 the SDC embarked on a programme of derelict land reclamation, clearance and infrastructure works to open up sites (see Table 7.2). As well as the recession, this change in emphasis also reflected a growing awareness about the extent of land pollution and the full costs of reclaiming such land and bringing it forward to the market (Hey *et al.*, 1997).

A significant feature of the property developments was the desire of the SDC to promote land uses which would diversify the Valley's economy and increase values in the area. Most notably this is reflected in the Meadowhall flagship, but there was also further retail, leisure and commercial developments and attempts to lure 'mega-projects' to the area, such as the Royal Armouries Museum which went to Leeds. Much of this diversification fitted in with the policies of the City Council, which was at the same time constructing two major sports/leisure schemes in the Valley adjacent to the urban development area, the Sheffield Arena and Stadium (see Figure 7.1). However, the promotion of office projects in the Valley was more controversial since it was claimed that these sucked development and employment out of the city centre, only two miles from the heart of the Valley.

The attitude of the SDC towards local businesses was also to change over its lifetime. Not only was the CPO débâcle seen as an indication of a negative attitude towards local traditional industries, but early promotional images created by the SDC suggested that the steel processors and scrap dealers did not fit with the new vision of the Valley held by the SDC. However, over time, the significance of many of these businesses to the economy of the Valley and the city, the international competitiveness of some of the leading companies, and the level of employment in this sector all contributed to a shift in attitude (Crocker, 1997; Raco, 1997; SDC, 1993). The result was a programme of business support grants and other measures, such as support for the metals and materials based Sheffield Regional Technopole. In other directions the SDC was less active and grants, for example, to the voluntary sector were less than 1% of total expenditure. Funds were spent on environmental improvements which had wider community benefits, for example the Five Weirs Walk, and the intended rebuilding of Attercliffe as a social community (see Figure 7.1).

Overall, the operation of the SDC continued a drift towards a style of urban regeneration that included: site specific property flagship developments; a dependence on a much criticised concept of trickle down to secure wider benefits; and a pragmatic strategy based on no clear industrial or economic analysis (Dabinett, 1991). At winding up in 1997, the SDC claimed that it had attracted some £683 million of private sector investment in land and property to the area and some further £300 million of investment in plant and machinery. Some 18,000 jobs had been created and 5,360,000 square feet of new floorspace had been constructed (SDC, 1997). On winding up, 'the roles of the Corporation as both regeneration authority and local planning authority for the area were returned to Sheffield City Council' (SDC, 1997, p. 2). The City Council was to set up the Sheffield City Development Agency, as a single purpose unit within the local authority. Exit from the Valley by the SDC was not to emerge as an issue in the city. The principles of property-led regeneration and partnership forms of governance were well embedded in the approach to urban policy more generally. Furthermore, extensive negotiations between the SDC and developers or project leaders led to an exit strategy based on individual contractual arrangements, site by site, development by development. This contractual form of urban management was also to become a more general feature of urban governance in Sheffield in the late 1990s, as the City Council placed more services with private contractors.

Appraisal of urban policy in Sheffield

Commentators have identified a range of criteria for assessing urban policy (Hambleton and Thomas, 1995). Our appraisal is not intended to be definitive but instead aims to bring into focus particular areas of debate. The first looks at business involvement and partnership. In this section the question of control, accountability and private sector involvement in the partnership is examined. The second focuses on funding regeneration and attempts to analyse the relationship between private and public investment. Finally the

Figure 7.2 *Meadowhall site after clearance before redevelopment*

third section looks at employment growth and distribution, examining arguments about equity and additionality in employment creation.

Business involvement and partnership

One view held by central government that underpinned the UDC approach was that urban local authorities had misused their power by excluding business from policy considerations, and allowing agendas to be shaped by particular political or factional interests. The use of a 'quango model' for the UDC structure was to overcome this (Centre for Local Economic Strategies, 1992). The Board of the SDC was appointed by central government and although three Sheffield councillors served on it, the majority of members were local businessmen. Many of these people were already involved in other public/private bodies set up to manage urban regeneration in the city. There

Figure 7.3 *Meadowhall*

was considerable overlap between membership of the Sheffield Training and Enterprise Council Board, Sheffield Partnerships Ltd, Universiade GB Ltd, Sheffield Science and Technology Park Boards, Sheffield 2000, SERC and more significantly the City Liaison Group (Strange, 1997). Therefore, what part did the establishment of the SDC play in changing the role of business interests in the formulation of urban policy in Sheffield?

Representation on SERC was by appointment and after 1988 the SDC became a member. Between 1986 and 1992 SERC attempted to become the lead organisation in pursuing urban regeneration. Unlike the SDC it had no funds or executive powers and was subordinate to the central decision making bodies of the constituent organisations. Perhaps as a consequence of this, but also as a result of the leadership role taken by some members of SERC, the agendas were dominated by issues and discussions which avoided conflict. Any disputes were seen as potential deterrents to inward investment, a key objective of the group in the late 1980s. Consensus was seen as a necessary prerequisite to a favourable image for the city, an image which was fundamentally based on the criteria of private capital and business. Therefore, the establishment of the SDC might be regarded as simply taking the influence of the business sector a stage further, rather than creating a radical departure in the city's politics. Whilst this might be consistent with the fragmentation of urban governance and new forms of urban leadership during the 1990s, it can ignore the important issues of accountability and power within these emerging partnerships. The formal establishment of agencies such as the SDC undeniably diminished the powers and resources available to local government, but the informal 'leadership' partnerships are also important in this context. Together, they have assimilated areas of governance once largely the preserve of local authorities and the institutional and individual membership is not subject to formal electoral processes. Certainly the 'coalition framework which emerged in Sheffield was one which produced an enhanced role for business interests, but which also allowed the local authority a central position in their mediation' (Strange, 1997 p. 15).

But has this drift to public-private partnerships in urban strategic leadership constrained the policy agenda? The situation in Sheffield which had emerged by the mid-1990s provides an interesting perspective. When SERC was originally created in 1987, the emphasis was firmly placed on economic and physical development, image and inward investment. Yet by 1995, the City Liaison Group argued that regeneration meant creating a culture of optimism, maximising opportunity and choice and promoting jobs, social integration, participatory self-efficiency, self-belief and diversity (Sheffield City Liaison Group, 1995). These aims would certainly not have figured in this way in documents emerging from the earlier partnerships, and this is not the development dominated thinking of growth coalitions. Instead it reflects a much more imaginative and mature perspective on urban regeneration and its social, as well as its economic, dimensions (Dabinett and Lawless, 1998). What is less clear is the role that the SDC might have played in shaping these transitory perspectives.

An observer is likely to highlight the significance of the SDC, and in particular Board members within the overall governance and leadership of the city, but the extent to which any mutual transformation occurred within these relationships is difficult to establish (Mackintosh, 1992). The overriding impression is that a private market, property led, economic deterministic approach was pursued in the Valley, despite the SDC's support of the wider SCLG objectives for the city.

A final comment on the nature of partnership concerns the wider interpretation that can be placed on these events in terms of central government objectives. It is of interest that the early establishment of a public/private body in Sheffield was not directly supported by central government. Despite the contrary views of SERC, a government-appointed UDC was seen as necessary. This might represent some expression of central control over local autonomy, and the SDC had very few delegated powers and had to seek approval from the DoE on many decisions. Alternatively the events might be interpreted as an attempt to split the local authority-led partnership in order to marginalise the role of the council even further. This later view was reinforced by the failure of the first City Challenge bid, the previous inability to secure central government funds for the WSG, and the subsequent success in securing substantial SRB funds once the partnerships in the city were reconfigured under the auspices of the SCLG. Therefore, the Sheffield experience can be seen to mirror the national transition in the changing role of the business sector in urban policy, but perhaps more significantly is also a commentary on the changing relation between central and local government.

Funding regeneration

The physical decay of the inner city is inherently linked to patterns of new private capital investment and the maintenance of existing capital stock. The main aim of the UDCs has been to bring about urban regeneration. In practice the vehicle for this has been property development. Although not returning to the physical determinism of planning in the 1950s and 1960s, this approach has placed private investment at the centre of urban policy in the last decade. In Sheffield private investment has been encouraged through partnership, and in particular the promotion of the city as an economically competitive location for investment, through the property market led policies of the UDC and through subsidies in the form of Derelict Land and City Grants.

Private sector investment and its relationship to public sector spending through the notion of leverage was used by central government throughout the 1980s as a way of judging the success of urban policy. However, despite the commitment from the City Council in Sheffield to partnership, it is in the delivery of private capital investment within local authority projects that this approach most obviously failed. Most major initiatives ended up being paid for by the public sector. This was most dramatically illustrated by the staging of the WSG in 1991 where the private sector contribution was minimal. This might reflect the nature of the partnership in the city. It was based on local political consensus and joint promotion activities rather than a property or regeneration company. It might more fundamentally reflect the lack of competitive advantage of the city to mobile capital at that point of time, or the weakness of the local property markets. By contrast, at the time of winding up, the SDC claimed that some £600 million of private sector investment had been attracted to its operational area, though this included Meadowhall (some £250 million of investment). Furthermore, Sheffield City Council indicated that the total value of public and private projects awaiting construction in the city by March 1996 was £580 million. These are considerable achievements when compared to the recession years of the mid-1980s.

The contribution of private sector capital in urban policy and urban regeneration initiatives raises questions about the basis of private sector involvement. For private sector investment in development proposals, the main aim is to lever out public sector money in the form of grants and infrastructure work in order to increase returns on capital and reduce the risk. The question of who is levering whom is clearly crucial. One outcome from the policy towards urban regeneration pursued in the 1980s was a tendency to go for bigger and bigger solutions, perhaps exemplified by the WSG and SDC flagship proposals. The inverse of this was less activity in promoting the fine grained and small scale projects, that, taken together, could have made a difference in deprived areas of the city. One major scheme that illustrates this market-led approach was the Canal Basin, later to be re-named Victoria Quays. The derelict Canal Basin had been an environmental problem for decades prior to the SDC's incorporation, but it provided an ideal waterfront development opportunity, so popular amongst urban flagships of this period.

Shearwater were the original winners of a City Council backed competition for the site, but after 1988, under the SDC, they negotiated a £10 million City Grant for the scheme which included retail, housing, office, hotel and leisure development. Shearwater pulled out of this scheme in 1990 and a new competition was announced which was won by Norwest Holst. The new scheme had a higher office content with less retail and leisure, and was awarded an £8 million City Grant but no work had started by 1992. In 1993 the SDC took a fresh look at this site and a more interventionist approach, with more direct funding to core elements of the site. According to the SDC (1996) – 'The success of this approach is evident, with all the listed buildings now refurbished; a new multi storey car park on the site; new offices already occupied by Nabarro Nathanson; a four star hotel currently under construction; and all parts of the site subject to development agreements progressing at a rapid pace' (p. 10). Therefore, proactive planning with the 'public' sector acting as the key enabler released the development potential of this site. Previously, the need to scale-up the development to ensure that extensive site costs could be met as well as the required private rates of returns, had led to a series of unviable proposals.

Employment growth and equity

In February 1997, nearing winding-up, the SDC claimed to have achieved 18,000 jobs, including those in Meadowhall. The SDC had been set a target of 20,000 jobs. However, they also claimed that the 'conditions for self-sustaining regeneration have been put in place and we are confident that in the immediate future, as a direct result of the Corporation's work a further 6,400 jobs will be attracted to the area' (SDC, 1997, p. 7). It is, therefore, interesting to reflect that surveys undertaken for the SDC within the Valley indicate that employment rose from 17,900 in 1988 to 25,700 by 1997, an increase of nearly 8,000 (see Table 7.3). Whatever statistical anomalies that may exist, any job creation by the SDC has clearly maintained employment levels and off-set further job losses as much as it has been able to create additional employment in the Valley.

In 1997 the Lower Don Valley contained some 1,100 firms involved in a variety of sectors (Crocker, 1997). The two sectors most represented were distribution, some 600 businesses, and engineering, some 200 businesses (see Table 7.4). Similarly, the main element in the growth of jobs has also been in these two sectors. The structure of the Valley's industry has changed over the life of the SDC. Employment losses have still continued in some sectors, most notably metal manufacture, but the economy has diversified with growth in transport, distribution, finance and other manufacturing activities. The largest gain has been in distribution, largely, but not entirely, as a result of Meadowhall. The 'steel valley' is now the 'leisure valley', as retailing has become a dominant employer, with its share of employment rising from 15% of the workforce in 1990 to 38% in 1997.

One trend these surveys highlighted was the increasingly skilled nature of the workforce in the Lower Don Valley over the past decade (Crocker, 1997). The proportion of people in professional and technologist occupations rose from 5% of those employed in 1990 to 13% by 1997. The proportion of skilled and semi-skilled manual workers has risen, but the proportion of manual workers has fallen significantly. This is likely to be the result of firms having to improve their levels of skills in order to compete in national and international markets (Crocker, 1997). Despite the opening of Meadowhall, the main category of employment in the Valley remained male and full-time. Unlike the 'traditional' part of the Valley, employment in Meadowhall is 77% female, with 62% working less than 30 hours per week and 8% employed on a casual basis. The 'traditional' Valley also continues to be dominated by small firms of fewer than ten employees. Only three companies employed more than 750 people – Avesta (stainless steel manufacturer); Forgemasters (a group of seven businesses); and Tinsley Wire (Belgian owned corporation). Total

Table 7.3: Employment in the Lower Don Valley 1988–1997

(thousands)	1988	1990	1991	1993	1994	1995	1996	1997
Jobs in Meadowhall	NA	NA	4.9	5.3	5.9	6.4	6.2	6.3
Jobs in rest of LDV	17.9	18.8	17.9	16.8	17.2	16.6	18.8	19.4
All jobs	17.9	18.8	22.8	22.1	23.1	25.0	24.9	25.7

Source: SDC Business Surveys (Crocker, 1997)

Table 7.4: Business sector change in the Lower Don Valley 1990–1997

SIC Division	No.of firms 1990	No. of firms 1997	No. of jobs 1990	No. of jobs 1997	Change in jobs 1990-1997
1. Energy/water	6	1	1,101	52	-1,049
2. Metal mnfr	40	36	4,274	1,941	-2,333
3. Engineering	199	208	4,904	7,925	+3,021
4. Other mnfr	40	45	733	895	+162
5. Construction	15	45	1,625	1,285	-340
6. Distribution	275	604	2,779	9,854	+7,075
7. Transport	42	52	616	1,156	+540
8. Finance	50	78	713	1,606	+894
9. Other services	27	61	1435	1,023	-412

Source: SDC Business Surveys (Crocker, 1997)

employment in these three companies was nearly the equivalent of the largest single employer in Meadowhall, Sainsbury's Savacentre. This picture therefore portrays an economy in considerable flux, as the nature and levels of employment reflect the continuing processes of industrial restructuring and urban regeneration.

The property-led approach to regeneration was supposedly to provide links to the well-being of all urban residents, an argument crystallised in the notion of 'trickle-down'. However, evidence would suggest for many individuals and households in the more deprived neighbourhoods, a range of institutional, cultural and labour market barriers constrained already limited potential employment emerging from the SDC projects (Lawless, 1995). Studies in particular highlight the severe situation faced by unemployed males, and the Sheffield TEC also argued that job losses continued in the mid-1990s due to public expenditure constraints, the closure of Sheffield based branches of local companies, and the continuing difficult trading circumstances in steel and engineering. The official unemployment statistics also indicated a city which had not yet fully 'turned the corner' and embarked on a self-sustaining growth trajectory. In February 1996, the seasonally adjusted rate for unemployment in the UK was 7.9% and in the region of Yorkshire and the Humber it was 8.9%. The equivalent figure for the Sheffield travel-to-work-area was 10.3%. It may be argued that the benefits of regeneration in the city generally would most likely affect intra-urban variations in unemployment. The SDC and other projects implemented in this decade were located within or close to the inner city areas of Sheffield. It might be anticipated that some of the benefits would leak into surrounding areas. However, the marked socio-economic patterns of poverty within the city seem to have been accentuated if anything (Dabinett and Lawless, 1998). In effect, local people were not generally benefiting from the new, local, employment opportunities in retail or commercial activities (Lawless, 1995).

An urban policy as if people matter

Those directly involved with the SDC have been amongst the first to claim its success – 'While there can be no room for complacency, the Corporation has put in place the essential conditions for the regeneration process to continue in the Lower Don Valley' (SDC, 1997, p. 9). The legacy it regarded as leaving the city is probably best expressed in the four key lessons for urban regeneration it chooses itself to highlight from its experience –

- the need to assemble sites, by Compulsory Purchase Powers if necessary;
- the need for a transport infrastructure;
- the need to work with partners, both public and private; and
- the need for a focused and holistic approach to marketing, always placing the customer first (SDC, 1997).

Without doubt the Lower Don Valley has been transformed. Meadowhall is a very successful shopping centre, served well by both private and public transport links. The same links serve an ever emerging group of leisure facilities – an arena, a stadium, clubs and nightspots, hotels, retail centres and family pubs, car showrooms and multi-plex cinemas etc. In addition, significant new investment and activities have been attracted – an Abbey National sharetrading centre, back offices of the Halifax building society, offices of London-based solicitors Nabarro Nathanson, a Freemans mail-order centre etc. The Canal Basin, now named Victoria Quays, and the city airport are fully or partially developed, many of the large sites such as Atlas are partially or fully occupied by a mix of office, business and industrial uses. Whilst some of these developments have been occupied by companies that have moved within Sheffield, such as Neills Tools, this has often provided an opportunity for these companies to secure modern and efficient premises. The Valley appears to have been modernised, and its regeneration begun.

However, in 1993 we asserted that the main purpose of urban policy should be to improve the quality of life of people who live in cities, and in particular, those who experience disadvantage. All projects funded out of the public sector resources made available through urban policy should be able to illustrate clear and substantial benefits to the disadvantaged groups in the city region. The failure of the SDC to adequately fulfil such criteria is illustrated by the foregoing analysis of employment change. This failure has been based on the false analysis of the 'trickle-down effect', which assumes that major developments will create benefits that, in time, will provide opportunities for the urban populace to take advantage of. This, in turn, led to urban policy becoming fixated on site-specific projects, an approach that dominates the planning framework of the SDC. Too much money from the urban policy purse went into property-based flagship projects. Such projects were often a reaction to malfunctions in local property markets, which perhaps should have been dealt with by other measures. The UDCs were often seeking to fill gaps left by inadequate regional and national policies. The recent proposals for regional development agencies in England supports this case.

We also argued in 1993 for an enabling and innovative urban policy, which should not be dependent on property cycles and should not be the preserve of powerful people who can operate at the flagship scale. In essence, our view of an urban policy for people was one smaller in ambition, broader in scope, and more neighbourhood based. This approach would see more initiatives such as law centres, adult education projects, debt advice agencies, community and development trusts. Above all it would deliver policies through the involvement of people and communities rather than simply seeking tokenist community representation to support transforming visions. The introduction of SRB in 1994, and the successful attraction of these challenge funds to Sheffield therefore represents a pertinent test for this argument, since it is based on smaller scale, community-based and holistic programmes and to date has sought to consult extensively with local stakeholders.

However, those who still work to regenerate the Lower Don Valley and seek to bring about a more equitable socio-economic structure in the city do so within a framework of urban governance which clearly reflects the legacy of the SDC:

- the use of single purpose agencies to lead physical regeneration (the Sheffield City Development Agency) and urban regeneration (the Sheffield SRB Partnership);
- an approach to urban renewal based on partnerships with the private sector and property-led flagships, as illustrated by the redevelopment of the 'Heart of the City' using National Lottery funds;
- new forms of contractual arrangements and the fragmentation in urban governance with previous local authority services being contracted out to private and charitable bodies for provision and management; and
- wider social and equity goals being addressed by increasingly targeted actions.

Ultimately, the central messages for those seeking to succeed in the regeneration game over the period studied were partnership and privatism. By the early 1990s it was clear that the key actors and agencies in Sheffield – the local authority, the Sheffield TEC, the SDC, the Chamber and Cutlers Company, and the Universities were eager to enhance the process of regeneration. Hence, because SERC had come to be perceived as an ineffective executive, a new and much more powerful organisation, the SCLG, emerged. But it is not possible to locate the SCLG solely within the context of Sheffield. SCLG reflected the national government's determination to instil business attitudes and aspirations into regeneration programmes, and urban governance more generally. A process clearly reinforced and strengthened by the SDC, and perhaps the SDC also effectively initiated this process within Sheffield. Thus we saw the first of the failed City Challenge bids in particular being driven from the 'local authority bunker'. In contrast, the successful SRB bids emerged from a policy environment deeply rooted in partnership with explicit roles for the private sector and the community. The relationship between the SCLG, SDC and this success story is indeed an intriguing one. Certainly the notion of Peck and Tickell (1994) that partnerships in urban regeneration within the UK are best regarded as 'grant' rather than 'growth' coalitions very much reflects the Sheffield experience, and the culture of pragmatism which has evolved within the city.

The SDC has been successful in regenerating the Lower Don Valley. We are unable to answer the difficult question if this was the only or the best way to achieve the same ends. We are convinced that the approach nested in a wider approach to urban policy in the 1980s which was inappropriate and inadequate. The current implementation of SRB and use of EU funds such as URBAN in the city will provide an opportunity to explore the notion of an 'urban policy for people'. It will, though, take at least a further ten years to know if these recent regeneration activities will lead to wider empowerment, a sustainable economic future, and a better quality of life for those residents of the city who have suffered such widespread and continued disadvantage during the 1980s and 1990s. To date, the Sheffield SRB Partnership has targeted some of the most deprived areas and excluded communities in the City, and has programmes to support community enterprise, ethnic minorities, young people and community safety and a Sheffield Capacity Building Initiative, the last being matched with EU funds. Thus, it would appear that as the largest single recipient of SRB Challenge Funding, Sheffield has a genuine opportunity to improve the quality of life for people in the City.

Further reading

Sheffield provides an interesting example of industrial restructuring and institutional change. The latter has been extensively studied, in particular by local academics such as Lawless (1990; 1991) and Seyd (1990), and in more specific research, for example Raco (1997) and Strange (1997). The industrial history and dramatic decline in traditional activities also provides useful case study material, well recorded in the writings of Dabinett (1995), Tweedale (1995) and Hey, Olive and Liddament (1997).

8

'Out of touch, out of place, out of time': a valediction for Bristol Development Corporation

NICK OATLEY AND ANDREW MAY

Introduction

Bristol Development Corporation (BDC) was announced as one of the third phase Urban Development Corporations in 1987, although its official opening was delayed by 18 months due to a petition by Bristol City Council to the House of Lords opposing its establishment. Bristol's single-purpose, centrally appointed body, was given a brief to regenerate a central area of the city using extensive powers, public and privately generated funding focusing explicitly on property-led, market oriented strategies. Politically, the government used BDC to bypass the perceived cumbersome, inflexible and commercially insensitive statutory local authorities. Bristol City Council in particular was considered to be anti-development, inflexible and too demanding with respect to planning applications. In this context in Bristol, as elsewhere, the UDC was used to redefine the interface between the public and private sectors at the local level and to undermine existing cultures of municipalist institutional governance.

This chapter reflects on the legacy of BDC. It revisits the progress made by the Corporation documented in Oatley (1993) and examines the key achievements of the Corporation. The chapter critically explores the political and institutional complexities of wind-up and exit, giving particular consideration to problems that arose as a result of the implementation of scant central government guidance on exit by a local organisation that had failed to embed itself in local policy networks. The conclusion draws out the lessons to be learnt from the turbulent experience of the Development Corporation in Bristol.

Bristol Development Corporation – key achievements

BDC, one of the third generation of UDCs, was established in January 1989 after the City Council had petitioned Parliament to object to the unwarranted intrusion into democratic local government planning and development powers (see Table 8.1). BDC's vision for the 850 acres of land lying to the south east of the city centre was to transform the image of Bristol to one of the great cities of Europe by encouraging 'a dazzling, modern style of architecture' which would reflect the uses of financial services, hi-technology industry and medical technology supplies industries that it sought to attract (BDC, 1989). BDC saw itself as spearheading the physical and economic regeneration of the area through the attraction of high value added economic activities (Figure 8.1 shows the location of Bristol Development Area).

When BDC was wound up in December 1995 and closed down in March 1996, after seven years and two months of turbulent operation during which many of its plans had been actively opposed by the local authorities, its achievements had been more modest. It is widely agreed that its dream of transforming Bristol into a great European city remains a long way short of fulfilment. The National Audit Office's (1997, p. 49) report on the wind up of Leeds and Bristol Development Corporations stated that 'The Corporation (Bristol) only partially fulfilled the Department's expectations with regard to its regeneration statement'. Robson *et al.*'s (1998a) report, commissioned by

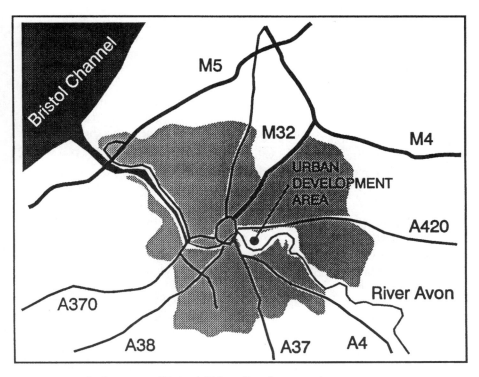

Figure 8.1 *The location of Bristol Urban Development Area*

Table 8.1: Chronology of events

Dates	Key Events
December 1987	Secretary of State for the Environment announces decision to establish Corporation
May 1988	Bristol City Council petition House of Lords against the establishment of the Corporation
November 1988	House of Lord's rejects Bristol City Council's petition but reduces the size of the urban development area
January 1989	Area and Constitution Order comes into force
February 1989	Corporation becomes local planning authority
November 1991	Bristol Development Corporation's life extended for one year
March 1994	Bristol Development Corporation's life extended for a further year
December 1995	Planning function Order removes planning powers
December 1995	Property Rights Order transfers residual assets and liabilities to the Secretary of State for the Environment
January 1996	Area and Constitution Order revokes designation of Urban Development Area
January 1996	Dissolution Order ceases Urban Development Corporation operations – except to close down operations ('wind down')
March 1996	Corporation closes

Source: National Audit Office (1997)

the Department of the Environment, Transport and the Regions (DETR), is also critical of Bristol's performance. Its main achievement was the construction of the Spine Road which attracted retail and leisure use at Avon Meads and Castle Court. In spite of this development, BDC was unable to realise its goal of attracting high value added industries and failed to develop the most attractive site in the Urban Development Area (Quay Point, or Temple Quay as it is now known). Unlike other UDCs established in the third wave of designations, BDC did not exert any significant local political influence by embedding itself in existing policy networks. Indeed, it purposely positioned itself outside of such networks even during the active years of partnership formation which characterised the local political scene from 1992 onwards. Consequently, its impact on the nature of governance in the area has been negligible.

The Corporation adopted three main strategies to promote regeneration. The first involved the building of a new Spine Road and associated minor road improvements to overcome access difficulties created by low bridges, railway lines and waterways. The second involved buying up selected blocks of land and selling consolidated sites to developers (e.g. Marsh Junction). The third strategy involved the active encouragement of landowners to bring

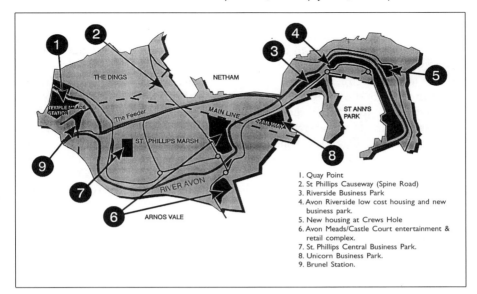

THE DINGS

NETHAM

TEMPLE MEADS STATION

The Feeder

MAIN LINE

RAILWAY

ST ANN'S PARK

ST PHILLIPS MARSH

RIVER AVON

ARNOS VALE

1. Quay Point
2. St Phillips Causeway (Spine Road)
3. Riverside Business Park
4. Avon Riverside low cost housing and new business park.
5. New housing at Crews Hole
6. Avon Meads/Castle Court entertainment & retail complex.
7. St. Phillips Central Business Park.
8. Unicorn Business Park.
9. Brunel Station.

Figure 8.2 *Key sites in Bristol's Urban Development Area*

forward land for development (St Annes). Figure 8.2 identifies the key sites within the Urban Development Area.

In spite of opposition by the City Council to the scheme, involving the displacement of over 54 companies and 600 jobs by the Compulsory Purchase Orders, the Corporation built St Philips Causeway (Spine Road), an elevated two kilometre dual carriageway, intended to open up an area with poor access and to take traffic away from the congested Temple Meads area of the city. The Corporation also reconstructed and upgraded some roads in other parts of the area, namely the Victoria and Albert Road system around St Philips Central Business Park, a new 2.8 hectare development (12,300 sq m). Although the Spine Road was built for £49m, £6m less than the Department's approved budget, there are disputes over the ongoing maintenance costs. Although the Corporation signed an agreement with Avon County Council, the then local highway authority, in May 1995 to adopt St Philips Causeway, Bristol City Council, which is now the highway authority, has raised concerns about a number of outstanding defects and the adequacy of the initial specifications which have implications for ongoing maintenance costs. This has meant that Bristol has not adopted the road and the matter has been subject to ongoing discussions between the Department and the City Council regarding future public sector costs.

St Philips Causeway opened up the Marsh Junction area which was developed into a major leisure and retail complex costing £50m (the Avon Meads and Castle Court developments). The area includes a Showcase Cinema (3,500 seater, 14 screen multiplex), 26 lane Hollywood Bowl, seven retail warehouse sheds, and six restaurants, a car dealership and a Sainsbury's cash and carry members club (which took over from an unsuccessful venture by the American bulk buy retail chain, Cargo Club). Because of the nature of the uses and the 'strip mall' feel of the design, this area became known as Little America. Although a popular leisure venue, this mix of uses is very different from that originally envisaged by the Corporation. Its original aspiration was for high value-added uses such as medical instruments manufacturers.

Further signs of under-achievement are evident in the Temple Meads area. The Temple Meads and Kingsley Village area was identified in the Corporation's `Vision for Bristol' as a potential gateway to the City, with the opportunity to achieve comprehensive mixed-use regeneration. The Development Strategy for the area included 97,000 sq m of office space, 10,000 sq m of retail use, 20,000 sq m of leisure use, 200 units of housing, and a 250 bed hotel. This formed the basis of a Compulsory Purchase Order which was challenged at a Public Inquiry in 1991, and granted in the High Court in 1992. In March 1994 the Secretary of State extended the Corporation's life for the second time by another year to December 1995, so that it could dispose of its flagship development site which became known as Temple Quay. In spite of this area being a key development opportunity the Corporation was unable to achieve any development on the site or to sell the site on to a commercial developer. As a result, the Department required the Corporation to transfer all rights to English Partnerships so that it could carry forward efforts to dispose of the site. The Corporation transferred Quay Point to English Partnerships for £6.9m in December 1995. Since the transfer, English Partnerships has continued to negotiate a sale of the site. Infrastructure works began on the site in April 1998.

In spite of early difficulties in finding a developer for the area, the Corporation did eventually secure the development of a new waterside urban village in the Avon Valley consisting of 600 homes, shops, a doctor's surgery and a nursery. This development was achieved on a difficult, polluted site that had been vacant for ten years. Although the Corporation was involved in creating new communities and was located in close proximity to some of the poorest areas in the city (e.g. St Pauls and Easton) it supported only a limited range of community initiatives. The National Audit Office Report (1993) on the achievements of the Second and Third Generation UDCs showed that BDC spent the least on supporting local community activities of all the second and third generation UDCs, both in terms of amount (£40,000 up to 31 March 1992) and as a percentage of its total expenditure (0.1%).

Table 8.2 summarises the main aspirations of the BDC for the key sites in the area and what was finally achieved.

Table 8.2 The main aspirations as contained in BDC strategy documents compared with what has been achieved on the ground

Site/Project	Initial Aims	Achievements
Quay Point	Mixed development of offices, retail and housing. The 'jewel in the crown' of the area	Transferred to English Partnerships and still awaiting development
St Philips Causeway/ The Spine Road	Road linking M32 and the A4. Central part of BDC strategy to open up St Philips Marsh and to bypass congested Temple Meads area	Completed ahead of schedule in July 1994 at a cost of £49, £6m less than the Department's approved budget. On wind up BDC unable to transfer the road due to defects. 54 companies and 600 jobs relocated
Avon Riverside (including Riverside Business Park) at Crew's Hole	Housing scheme on both sides of the river including plans for a business park	New waterside urban village created including shops, a doctor's surgery and a nursery, restoration of jetties and a new business park. 600 homes complete and sold. 1,100 homes with planning permission
Avon Meads and Castle Court	Originally planned to develop a business park with high value added uses e.g. medical industries, R+D	Avon Meads and Castle Court entertainment and retail complex (Bristol's 'Little America'). Yielded £13m capital receipts for Corporation
St Philips Central Business Park	Improve road infrastructure	New road links to the Spine Road
Brunel Station	Environmental improvements	Improvements to Station approach
Avon Weir	Building of a weir to create better environment	Not developed

Source: BDC strategy documents

The political and institutional complexities of wind-up

The difficulty of exit for Britain's development corporations has been acute (also, see Chapter 1). This has been because of their financial dependence on rising land and property markets which has rendered them vulnerable to economic downturn but also because insufficient attention has been given to the highly political nature of exit. While 'exit strategies' were widely produced these were not able to address adequately the complicated issues arising from exit and succession. Official guidance on exit arrangements and their monitoring has concentrated on a narrow range of issues – 'leaving the shop tidy' and leaving the political brokering of agreements to local circumstances.

Reliance of the development corporations on capital receipts from land and property meant that their financial and project planning were badly hit by the property recession of the late 1980s and early 1990s. By August 1993 the National Audit Office (NAO, 1993 p. 8) noted that 'UDC's have taken a number of steps to ensure orderly wind-up but the slowing down of development and uncertainty about wind-up dates have made it difficult to assess . . . the extent of potential problems'. 'Unfinished business' has been the inevitable result when sites fell in value and projects were either delayed or cancelled. In Bristol the BDC's life was extended twice due to the delay in bringing a key site (Quay Point) to completion while another large-scale project (Avon Weir) was abandoned. The tidy exit sketched out for example in the NAO report of 1993 looked very difficult to achieve.

As a result, with eight of Britain's DCs due to wind up between December 1995 and March 1998, a new issue of 'incomplete regeneration' began to emerge. This manifested itself in different ways. First there was the problem of individual development projects left incomplete or not achieved and secondly the wider problem of strategies not fully implemented with inevitable succession problems and lack of continuity from successor organisations. A third range of issues, ostensibly more straightforward, is also present: the hand over of maintenance or management costs of completed projects, particularly to local authorities. These particular problems seem almost universal – not surprising given local authority budget problems. In Bristol's case Local Government Re-organisation exacerbated problems in reaching agreement, for example over the BDC spine road, in that the successor highway authority (Avon County Council) was itself due to disappear only three months after the BDC!

Interestingly, as Chapter 1 also indicates, the issues of wind-up or exit are nowhere addressed in the primary legislation (Local Government Planning and Land Act 1980) and where guidance was issued later it envisages exit as an orderly, rational process with sufficient time, energy and organisational capacity for agreement to be reached. Rather, exit is essentially a political process in that it deals with difficult, even intractable, issues about power relationships, differing cultures and above all differing interests.

In the light of this, 'model' arrangements such as those on offer in the formal guidance are of limited usefulness. Rather, examining exit as being about the interplay of different interests (as examined below) may enable us to better understand the process. For example, as wind-up approached BDC gave priority to two main objectives: to secure a positive verdict on its tenure and to ensure its flagship project was committed. Given the widespread problem of unfinished business Development Corporations will try to commit their successors to fulfil their plans. Planning consent was granted for Quay Point in

Bristol, for example, against the request of both Bristol City and Avon County Council, just weeks before the BDC wind-up. Ambitions to 'leave something behind' may lead to tombstone planning – that is, to memorial activities.

The local authority perspective will be very different, being driven by genuinely differing interests. Given the financial difficulties of most UK local authorities, the need to minimise or eliminate liabilities inherited from the departing development corporation will be paramount. Many completed capital projects with maintenance implications may well be given a 'dowry' by the departing development corporation . However, reaching agreement about what is reasonable may be straightforward about say highway verge grass cutting but very contentious where maintenance costs are hard to predict. In the case of Bristol Spine Road ('St Philips Causeway') agreement was never reached due to the unorthodox construction methods used – and ownership and responsibility for the structure are still unresolved. The local authority too will seek to normalise the policy regime, to re-assert control (e.g. the recovery of planning control) over an area which has had a special regime while it was the Urban Development Area (UDA). Concentration on the UDA or on individual flagships by the UDCs is notoriously un-strategic and the local authority, with a more strategic point of view, will be anxious to assert it.

The DETR as overseers of exit will have a complex set of interests. As sponsors of the development corporation they too will seek a positive view of the record of the corporation. In addition they will seek an orderly exit and the future relationship with the local authority may well be something that preoccupies at least the local office during the transition. Certainly the NAO's 1997 report on the wind-up of the Leeds and Bristol development corporations makes clear that the relationship between the DETR/Regional Office and the development corporation during wind-up was a private two-way dialogue aimed at minimising disruption and reconciling differences. Interestingly, there is a little sign of the DETR adopting much of a monitoring or scrutiny role: rather this has been seen as the province of the NAO or perhaps later Parliament itself.

The Treasury, concerned at the overall impact of the development corporations, may almost certainly have overseen the recovery of costs through land and property sales although direct evidence of this is hard to find. In the Bristol case for example a proposal was made shortly before the wind-up of the Corporation for the Exchequer to receive a share in later profits from development following the sale of land at Quay Point (National Audit Office, 1997 p. 39).

In some Urban Development Areas and notably in Bristol the presence of English Partnerships as a kind of successor body brought another set of interests to bear upon the exit process. English Partnerships, with its very particular post-development corporation, post-City Challenge, culture and changed expectations, has been well-placed to act as broker between the corporation legacy and the local authority. At other times, as a kind of surrogate for the local authority, emphasising partnership and co-operation and, at the same time, open to the more strategic viewpoint of the local authority, English Partnerships has had a crucial reconciling role in achieving policy stability and in improving relationships (see the section on Quay Point below). The aim has been to achieve a policy stability and to improve relationships. Uncertainty over Quay Point in Bristol, for example, has been reduced by English Partnerships as landowner and as founder and sponsor of the Harbourside project elsewhere in the city. By establishing its interests in both Quay Point and Harbourside, English Partnerships has been perceived as taking a balanced, less partisan, view of city centre development politics.

Exit, in much of the urban development area, was fairly straightforward: in large areas where Corporation activity had been minimal or was completed, exit was simply a matter of de-designation. Indeed Bristol City Council as the successor planning authority sought early de-designation of much of the urban development area in 1994, only to be rebuffed as the BDC proved quite tenacious over its development control powers.

Issues that arose later were in many places limited to debates about assets and liabilities. Bristol City Council, as the inheritor of the Development Corporation's completed projects (most notably the Spine Road), adopted a negotiating position of 'no net liability to local tax-payers'. The BDC response was to offer packages of funding to support maintenance or on-going revenue costs. Such a 'dowry' solution has been common in all development corporations and any disagreements have largely been about the adequacy of particular funding arrangements.

However, difficulty arose in Bristol over the definition of some liabilities. For example, Bristol City Council sought a contribution towards continued air quality monitoring of the effects of BDC highway building works; there were disagreements too about the lack of provision (e.g. through section 106 agreements) of certain 'goods'. Examples of this were children's play areas in new residential developments and funding for supported bus services to provide public transport to new leisure developments at Avon Meads (multi-screen cinema, fast food restaurants, bowling alley and edge-of-town retail outlets) which were criticised as being only accessible by car. Lack of any definition in government guidance on exit meant that the context for resolving disputes of this kind was unclear. In the environment of winding down of the development corporation, failure to reach agreement about such problems turned into a solution by default, as the development corporation simply sat out the time available to reach any agreement.

The Corporation sought to continue its regeneration activities until the last possible moment but this approach was at the expense of a sharper focus on the need to wind up the Corporation's affairs and meant that a significant number of tasks were left to be completed by the Department of the Environment. In May 1994 the Corporation produced its initial wind-up strategy which identified 50 tasks together with specified milestones. The DoE was concerned that this initial strategy did not provide enough detail to allow adequate monitoring to take place. The Corporation employed consultants Sir

William Halcrow and Partners to produce clearer documentation on progress in completing the wind-up tasks. In spite of more detailed monitoring on the progress of these tasks when the Corporation closed down in March 1996 it left over 100 tasks to the Department of the Environment to complete. The National Audit Office report (1997, para 3.47) noted that these tasks involved an estimated cost of up to £10.4m, only £5.5m of which was covered by the Corporation's cash surplus at the end of its life. Furthermore, the Department estimated that it would take until the end of 1997 to deal with the outstanding tasks, at a cost of some £677,000 in staff and consultancy fees. The wind-up was made more difficult because, in spite of uncertainty about whether tasks would be completed on time, the Corporation did not produce a documented risk assessment either to help its own management of wind-up or to prepare the Department for the likely tasks it would inherit.

The Department of the Environment anticipated that it would be able to complete the majority of inherited tasks by the end of 1997, although three tasks would take longer to resolve; the adoption of St Philips Causeway, the ongoing responsibility for the Avon jetties and 32 outstanding compulsory purchase order claims.

Quay Point

The process of exit and the issue of 'tombstone planning' are both illustrated by the recent history of Quay Point. Throughout its seven-year life BDC inevitably focused its efforts on four or five key projects although they were rarely identified as such. By far the largest, with the most potential impact on the city, was the site north of Temple Meads station variously known as Quay Point or – after 1995 – Temple Quay. The final two years of BDC's life were increasingly focused here on the impact of incomplete regeneration and the need to manage competing demands of tidy exit and recovery of land acquisition costs by selling the site. Energy focused increasingly on this 22 acre strategic site following a lengthy and difficult compulsory purchase order inquiry in the early 1990s. Key decisions about the future of the site were deeply influenced by wind-up and exit. For example, the extensions of life given to the BDC were widely perceived as being justified initially by the slow progress made in the face of a collapse in the land and property market (1991 extension) and then the need for more time to complete Quay Point (1994 extension).

The site, bounded by the inner circuit road which connects directly to the national motorway network, by Temple Meads itself and by water (an arm of the historic floating harbour) is both very accessible and clearly defined. Lying at the south-east edge of the city centre, it represents an obvious opportunity to extend the city centre uses and to integrate Temple Meads which, like many Victorian railway stations, is notoriously remote from the centre of the city (see Figure 8.3).

In 1989 the first planning consent granted by the BDC for the core site

Figure 8.3 *Temple Meads remote from the city centre*

of Quay Point was for about half a million square feet of B1 (offices) but only shortly afterwards, in 1990, a 'Planning Framework' was issued by BDC as part of the CPO enquiry evidence which envisaged substantial off-site infrastructure works. By 1994, with wind-up approaching, a radical shift of land use to retailing was made: a 300,000 square feet department store was mooted together with large scale leisure uses. This concept excited considerable opposition from both local authorities and from a virtually unanimous private sector in Bristol who all saw the scheme as another fragmentation of the city's retail core just at a moment when the need to support existing shopping at Broadmead against out-of-town threats was paramount. Finally, in May 1995, barely six months before exit, a more mixed land use was given planning approval. This included 1 million square feet of B1, up to 100,000 square feet of A3/A2 uses (restaurants, bars, cafes etc.), a 250 bed hotel, up to 200 residential units with 2,000 car parking spaces.

Pre-emptive final year decisions such as this are seen as undesirable in DoE guidance. However financial imperatives may well prove more powerful that anxieties about not fettering the discretion of those who follow the corporation. The National Audit Office review of the BDC records that in March 1994 (NAO, 1997 p. 39):

> The Secretary of State had extended the Corporation's life by one year primarily to allow more time to dispose of Quay Point (but) he insisted that contingency arrangements were necessary should a commercial sale not occur.

This 'switchback of policy changes' produced strategic conflict with Avon

County Council and also with Bristol, concerned about the strategic approach to the City Centre including tension with the City's own key site at Harbourside barely 3/4 mile away.

Despite the 1995 consent no attempts were made to implement the scheme until April 1998 when infrastructure works began on site. Ownership of the site (which had fallen dramatically in value) was in doubt as exit approached; DoE guidance on land and buildings held by DCs was that they should seek to negotiate disposal to other public sector bodies. There was little doubt in the Bristol case that this policy excluded the City Council and the site was sold to English Partnerships for £6.9m shortly before final exit of the BDC. English Partnerships has displayed a radically more conciliatory stance towards Bristol City Council (newly established since April 1996 as the unitary council for the city as well as the planning authority). The re-working of highway access into the site from Temple Gate within the terms of extant 1995 planning approval has now begun (April 1998). Temple Quay has been identified as a site for two important city-centre re-locations for Bristol and West Building Society and a local British Telecom re-grouping (negotiations over the BT re-location having been started by BDC).

Legacy of Bristol Development Corporation and emerging policy regimes

In addition to bringing about visible changes in run-down derelict areas, UDCs were also charged with the responsibility of contributing to changes in the socio-institutional landscape of the localities in which they were established. Central government began to roll back local government and redefine central-local relations with the introduction of the Local Government, Planning and Land Act 1980. Provisions for 'the establishment of corporations to regenerate urban areas', and to change local economic policy, land assembly and planning were all incorporated into this landmark Act (Local Government, Planning and Land Act 1980, chapter 65, Official Summary). UDCs were designed to show how the supposed energy, flair and drive of the private sector was more effective at generating economic development than the 'socialist' alternative involving public investment and state direction of resources. UDCs were meant to demonstrate a political lesson concerning the efficiency of the free market and the success of private enterprise versus the inefficiencies of democracy and state intervention (except that UDCs relied on massive state subsidies) (Duncan and Goodwin, 1988, p. 127). UDCs, then, were a central part of the government's attempts to redefine the frontier between the public and the private sector.

Consequently, although their success would be judged partly on their demonstration effect through their ability to bring about regeneration, it was also recognised that to have a lasting effect on the governance of the area, UDCs had to change the culture of local authorities and other local organisations more directly by redefining the frontier between the public and private sectors (Hansard, 13 September 1979 and Tom King quoted in *Local Government Chronicle*, 12 December 1980). This would depend on the extent to which UDCs could enlist the support of the private sector, develop working relationships with local authorities and other key local organisations and begin to exert an influence on their activities.

Both in terms of the demonstration effect and the wider processes of governance in the locality, the impact of the UDC in Bristol was negligible. There were five main reasons for the minimal impact of BDC on the locality. First, in terms of the demonstration effect, the achievements of the UDC were far from impressive (due largely to the state of the land and property market). The UDC had nothing it could point to that would persuade local authorities of the merits of its approach. The UDC even had difficulty in gaining the backing of sections of the private sector (the relations between the Chamber of Commerce and later the Bristol Chamber and Commerce Initiative were cool). They upset many of the small to medium sized businesses that were targeted for relocation due to regeneration plans. It had angered respected development companies and commercial interests that wanted to maintain the vitality and viability of Broadmead Shopping Centre (e.g. the Broadmead Traders Association and agents with city centre land interests), which was being threatened both by out-of-town shopping proposals and proposals by the UDC for large scale retailing at Temple Quay. Although this conflictual episode was short-lived and subsided when BDC withdrew its initial proposals for Quay Point, it demonstrated the lack of credibility of the organisation and the lack of links it had with broader communities of interest that would have provided an early indication of the inappropriateness of these proposals. Furthermore, its cavalier attitude towards the letting and management of consultancy contracts also frustrated a number of organisations (more detailed analyses of some of these issues can be found in Oatley's chapter on BDC in Imrie and Thomas, 1993b).

Second, the local authorities were influenced by the government's early rhetoric surrounding the establishment of UDCs, which presented them as a response to the failure of local government. Consequently local authorities were predisposed to opposition and against collaboration. This was particularly true of the local authorities in Bristol, where Bristol had objected to the establishment of the UDC in the House of Lords. This principled opposition never really subsided and was fuelled by a series of policy conflicts, personality and organisational culture clashes, and strategic political posturing of the Labour dominated Bristol City Council, which was vehemently opposed to the neo-liberal political agenda and policy instruments of central government designed to centralise power and diminish the role of local government (see Oatley, 1993).

The third factor was closely related to the above and was associated with the clash of organisational cultures and the personal leadership style of the Chief Executive of the BDC. From the outset, BDC and the local authorities were ideologically opposed. A war of words was conducted in the local press and although meetings were held between executive level UDC officers and Chief Officers and Councillors from Bristol and Avon County Councils and between less senior officers from both organisations, no reconciliation was brokered. BDC seemed determined to maintain its distance and seemed to cultivate its independence of existing policy networks. Rather than change local authorities' priorities and style of working, the UDC probably made local government in Bristol more resolute in adhering to its principles and policy priorities.

The fourth and fifth factors, arguably the most important, are closely related and involve the effects of the recession of 1989–1992 and the emergence of an alternative policy regime for regeneration in the form of City Challenge and subsequently the Single Regeneration Budget and its effects on the formation of local partnerships. The recession of 1989–1992 and the replacement of Margaret Thatcher by John Major as the leader of the Conservative government in 1991, led to a widespread re-evaluation of policy. In urban policy, Michael Heseltine, newly appointed as Secretary of State for the Environment in 1991, introduced City Challenge which shifted emphasis away from property-led, market driven solutions that undermined the role of local authorities towards a more integrated approach to social and economic problems that reintroduced an important role for local authorities. It was designed as an experiment to see whether a new competitive approach to funding in which localities had to bid for a limited pool of resources could overcome both the fragmentation and lack of integration in policy and a perceived 'dependency culture' on funding within local authorities (Oatley, 1998). Bristol was invited to bid in the first round and also submitted a bid in the second round but without success in either. This was a sobering experience for the City Council which had difficulty mobilising a limited and fragmented institutional capacity to present a convincing bid to government (Lambert and Oatley, 1995).

Bristol City Council was suspicious of the competitive nature of City Challenge leading to what some observers described as an ambivalent political commitment to the bids that were made (Malpass, 1994; Stewart, 1996a). Stewart (1996a, p. 126) observed that 'The imposition of the BDC and failure to win City Challenge twice reinforced what was perceived by some as the grudging nature of the authority's involvement with new coalitions and partnerships. Since 1992, however, public/private relations have been vastly improved. Stereotypes have been challenged, working together has become a habit rather than an experiment and public/private partnerships abound (many formalised in company legislation)'. With the Conservative Party returned for a fourth term of office in 1991 Bristol City Council grudgingly and pragmatically began to abandon its municipal socialist stance and began to take the issue of partnership more seriously than in the past. With the introduction of the Single Regeneration Budget and the Challenge Fund the City Challenge networks were resurrected and the Bristol

Regeneration Partnership was established, bringing together the Bristol Chamber of Commerce and Initiative, the TEC, representatives from Departments from within Bristol and Avon County Councils and a wide range of other public agencies (for example the police, Health Authority, and the universities). This Partnership body has contributed to the opening up of communication between different interests in the city, and led to the eventual establishment of a number of project partnerships between the public and private sectors.

It was the emergence of this new competitive policy regime, combined with a series of new appointments at Chief Executive and Directorate level in the Council and changes in the level of business activism in Bristol, that began to have a profound effect on the socio-institutional landscape within Bristol. The UDC was a by-stander in this process of change rather than an agent of change. The final impetus for change in the governance of Bristol was the worsening economic conditions in the period 1989–1992 which had a particularly severe impact on Bristol.

By the end of the 1980s Bristol found itself in the context of an uncertain economic future. In this context, tentative steps were taken to establish a greater degree of institutional thickness (Amin and Thrift, 1995) in Bristol leading to what Bassett (1995, p. 544) has called a period of partnerships and new business activism. The impact of the late 1980s and early 1990s recession sent a series of shock waves through the Bristol local economy, shattering the complacency that had existed in Bristol about the city's economic future. In particular, finance and business services shed 4% of their employment between 1989 and 1991 (Bassett, 1995 p. 544). The aerospace and related industries also experienced 23% reduction in employment for the same period partly as a result of defence cuts.

Unemployment in the Bristol area in general worsened during the early 1990s. In December 1989 Avon's unemployment rate (at 4.4%) was a quarter below the UK average. By May 1992 unemployment had risen to 9%, only fractionally below the UK rate, with acute levels of unemployment in specific Bristol wards. By 1993 unemployment rates in Bristol exceeded 13%.

The continuing fiscal constraints on Bristol through rate capping (1990–1992) combined with the prescription of the appropriate range of economic activities for local authorities in the Local Government and Housing Act of 1989 reduced the scope of activities and weakened the potential role of Bristol's economic development. At the same time central government was shifting resources and responsibility to the private sector for economic development/regeneration activities and encouraging the private sector to take a more active role in economic development efforts in localities. This meant that, increasingly, local authorities had little choice but to seek partnerships with the private sector to combine resources in the pursuit of job creation and growth.

In response to the economic problems of the city and the clear messages from government about changing the form of local governance, a new and more active business élite emerged in Bristol in the early 1990s. The Bristol Initiative (TBI), launched in 1991, was a Business Leadership team that set out to fill the gap created by the ineffectual Chamber of Commerce. The Initiative secured wide membership from ninety of the largest employers in Bristol, from local government, BDC and the Avon and Somerset Police and soon established itself as an influential player in a number of partnership activities such as the provision of social housing, the celebration of Bristol 97, and the formation of a Bristol Cultural Partnership. In 1993 TBI 'merged' with the Chamber of Commerce to create the Bristol Chamber of Commerce and Initiative, which secured its pivotal role in the increasing number of partnerships that were emerging in the Bristol sub-region (Snape and Stewart, 1995). Selective business interests began to coalesce around the Bristol Initiative, seeking to provide a forum for debate and a base for civic activity on the part of business.

There was a growing awareness among all key actors in the City of the increasing competition between cities for scarce mobile investment. A number of partnerships emerged in Bristol with the explicit purpose of developing a strategy and image for Bristol to compete with neighbouring areas and other cities in Britain and continental Europe (i.e. the Western Development Partnership, the West of England Initiative and West of England Economic Development Agency). These new organisations embodied the spirit of partnership that had developed in Bristol and the sub-region. This new mood was clearly expressed by the Chief Executive of the Western Development Partnership, who stated at the Press Conference of its launch that 'Partnership was the only key which will unlock us from the parochialism, rivalries, suspicions, complacency, and procrastinations that have hampered our region in the past, and given other regions, both in the UK and throughout Europe, a free hand to leave us standing' (Bassett, 1995, p. 545).

Meanwhile the BDC was becoming increasingly marginalised in this fast changing institutional landscape. The BDC has not been involved in BCCI networks. The Chief Executive of the BDC neither attended BCCI meetings nor participated in sub-groups. As the 1990s progressed, this time-limited body was increasingly seen as an irrelevance. Both public and private sector organisations looked forward to the time when English Partnerships would take over outstanding responsibilities of the UDC.

Conclusions

UDCs represented the high watermark of Thatcherite urban policy. They arose out of the socio-economic political context of the early 1980s and were imbued with the values of this period. They embodied the neo-liberal values of a post-municipalist world in which local authorities were by-passed, powers were centralised and shifted in favour of the private sector, and aggressive top-down, property-led regeneration was encouraged.

In the context of the 1990s, post-Thatcher, this approach seems wholly inappropriate. Not only does it seem overly confrontational and heavy handed but peculiarly one dimensional. In the current climate of the rediscovery of the importance of multi-sector partnerships and integrated regeneration strategies characteristic of City Challenge, Challenge Fund schemes and City Pride Prospectuses, the UDC approach with its single purpose focus on physical regeneration appears out of date and out of touch with current regeneration theory and practice (Oatley, 1998) (and with the model of the New Town Development Corporations on which UDCs were originally based). Even the emphasis on major road building as an effective means of regeneration, a distinctive feature of many Development Corporation strategies, has been seriously questioned by the publication of a radical report by the Parliamentary Standing Advisory Committee on Trunk Road Assessment (DETR/SACTRA, 1997) which concludes that road building, in some instances, does not always benefit the economy of an area and can damage the level of local economic activity.

An appreciation of the limitations of the Development Corporation approach to regeneration has grown and perceptions of how to address problems of run down urban areas have changed. A number of lessons have been learnt from the experience of UDCs in general and from BDC in particular.

Lack of a strategic integrated approach to regeneration

BDC lacked a long term strategic perspective and adopted a very narrow property-led approach to its task of regenerating a key part of central Bristol. There is now a strong consensus over the need for an integrated approach to regeneration, a principle first expressed in the 1977 White Paper (DoE, 1977). Contemporary practice in City Challenge, European Union and Single Regeneration Budget schemes show that integrated approaches stand a better chance of achieving lasting improvements and overcoming problems of social exclusion (Geddes, 1997; DoE, 1996a). Indeed, for Bristol it was partly the experience of participating in City Challenge and SRB bids which gave the impetus to new ways of partnership working within the city. The exclusive approach adopted by BDC served only to alienate both the public and private sectors within the city. It is now widely accepted that there is a need for inclusive approaches to regeneration building in the support of the local community and key local actors for lasting success.

The experience of the Development Corporations has demonstrated the limitations of the 'bricks and mortar' approach as a catalyst for regeneration (Turok, 1992; House of Commons, 1989; Robinson, 1997). Physical regeneration can play a part in regeneration, particularly if it is linked to creative schemes to link employment opportunities to the unemployed. Bristol's Harbourside development demonstrates what can be achieved if agreement can be reached over the importance of targeting employment opportunities. 'On-site', a local labour scheme, is aimed at targeting work and re-skilling opportunities arising from new development at the local unemployed. Jointly funded by

Bristol City Council, WESTEC, the Employment Service and English Partnerships the team is operating from its Harbourside base but is already working to expand city-wide including Quay Point. 'On-site' is also aiming to work with the windfall-tax funded New Deal.

Bristol's Urban Development Area, like many others, included strategic city centre sites of sub-regional importance. Although purporting to take a wider view, many of BDC's proposals conflicted with the policies in the Local Plan. In this sense, Bristol's experience highlights the inherent problems of establishing parallel planning systems, in which a time-limited quango can promote proposals that conflict with the policies of the local authority's Development Plan (Quay Point discussed above being a prime example). The pressure to achieve development within a short time scale and the temptation to promote flagship developments is not conducive to the achievement of long term strategic planning goals. One might have expected the Regional Office of the DoE to step in to try and resolve some of the conflicts, particularly over BDC's early plans for a large retail development at Temple Quay, at a time when Broadmead was under threat from plans to build an out-of-town regional shopping centre. The Regional Office of the DoE appeared unwilling or unable to broker an effective compromise over matters involving such strategic issues in Bristol. The combination of a newly appointed Director of the Regional Office of the DoE who wanted to avoid controversy and the ideological commitment of Ministers to UDCs provided a context in which backing was always given to the Development Corporation in order to assert the superiority of the UDC model over local authorities.

Pressure to demonstrate (quick) results

Regeneration is a long term process, requiring a long term commitment to strategic objectives. There was always the inbuilt temptation for Development Corporations to go for the dramatic, flagship development, attempting to bring about a transformation of the area overnight. However, quick fixes may not always be the best outcome for the locality. The ever present pressure to demonstrate results can often conflict with the long view. The problem of 'tombstone' planning (the desire to 'leave something behind') is a consequence of the pressure to demonstrate results. There is clearly a tension for time limited organisations who are working both to achieve results and plan their own exit. In Bristol more energy was devoted to securing results in the last months before wind up than to planning an exit which created problems and undermined the smooth handback of responsibility for the area to the local authority.

Appropriateness of the UDC model

The UDC model of regeneration was driven by an ideological motivation to restructure central-local government relations and the nature of governance, and to promote economic development based on addressing supply-side blockages in the development process of run-down areas. The experience of Quay Point in Bristol and the inability of the City Council to bring about development on the attractive Bond Street site (a strategic site next to the retail and office core of the city that was part of the original UDA only to be removed after the Petition, see Oatley, 1993) demonstrates the problems of relying on property-led regeneration regardless of whether it is under the auspices of a UDC, English Partnerships or the City Council.

Although there was considerable resistance to their actions in many other areas, in Bristol, tensions were acute partly because the UDC model of regeneration seemed less appropriate to the nature of the area. A single-purpose land regeneration agency dedicated to clearing and assembling land and promoting development may be able to generate a consensus over development more easily in an area with vast acres of dereliction. But Bristol City Council demonstrated that the area identified for Bristol's UDC did not contain vast tracts of derelict land and argued that such an approach was inappropriate. Throughout its life, it was always seen as an unacceptable and inappropriate imposition on the local area. BDC was perceived and experienced as a centrally imposed, political agency which paid little attention to the local democratically elected authority or to neighbouring communities. Whereas other UDCs, particularly some of the other second and third generation corporations, adopted a more pragmatic and conciliatory stance, partly as a result of becoming embedded in the local political and economic culture, BDC sought to maintain its independent and resolute identity.

Other UDCs which began to move beyond purely physical regeneration to more involvement in social, environmental and community projects and which engaged with local and strategic planning contexts proved to be more acceptable models for regeneration. This shift that took place in many UDCs was in keeping with the general shift in regeneration practice occurring in parallel recognising that trickle-down does not work and that the achievement of private sector development objectives is not incompatible with social goals. The importance of partnership between the local authority, private sector interests and community and voluntary organisations was also being learnt. A centralised quango with no local links was never going to be successful in achieving self-sustaining regeneration (Headicar, 1995, p. 5).

In the 1990s another model of regeneration emerged in stark contrast to UDCs. Post-local government reorganisation and post-Thatcher, local government lost much of its competence and capacity to act. Dedicated partnership agencies have been encouraged by central government policies such as City Challenge, the Single Regeneration Budget and City Pride and have filled the vacuum left by weakened local authorities. After initial difficulty in coming to terms with this new approach, Bristol City Council embraced the partnership ethic, even accepting an invitation to be among the second round of City Pride localities. Current policy regimes recognise the interconnectedness of urban problems and that to promote economic performance and competitiveness one also needs to address social disadvantage and exclusion and poor environmental quality. These values can be found in Bristol's SRB projects and underpin many of the partnership initiatives the authority is engaged in.

However, although the emphasis on property-led regeneration has receded, the promotion of entrepreneurialism and competitiveness remains. In this context, time-limited, semi-autonomous bodies with special powers and funding dedicated to achieving lasting impacts in difficult to regenerate areas are likely to always have a part to play in regeneration. The Bristol experience of UDCs has shown that for such agencies to work effectively they would have to be accountable, sensitive to the local political culture and embedded within existing partnership networks. They would have to pay regard to existing local and strategic policy frameworks and have a commitment to the current urban policy agenda. They would have to establish good working relationships with the local authority and ensure the conditions were secured for a smooth hand over of responsibilities on exit (CLES, 1990; 1992).

We can, perhaps, see examples of this approach to regeneration in the recent announcement by the government to set up special cross-departmental action zones in areas identified as suffering from acute poverty, similar to the American 'Urban Empowerment Zones' (Hambleton, 1995). In these areas the British government intends to establish local partnerships to channel multi-billion pound programmes to tackle the complex problems of rundown urban areas (*The Guardian*, 24 January 1998, p. 1). Such approaches, combined with national programmes to create employment and reform the welfare state, could release the potential of lasting regeneration in a way that the single purpose UDCs failed to do. They could build on the synergy of participating agencies and although they would have a clear area focus they would have a multi-purpose agenda. However, these approaches would still need to overcome, more successfully than was achieved by the UDCs, issues of local political tensions around policy priorities, accountability, participation and ownership if the conflicts of the 1980s are not to re-appear. The establishment of the Regional Development Agency in the South West will provide a stiff test for this new style, single-purpose quango with powers over and above elected local authorities.

Further reading

Bristol was one of the five related city studies carried out under the Economic and Social Research Council's Inner Cities Research Programme during the early 1980s. As a starting point *Sunbelt City? A Study of Economic Change in Britain's M4 Growth Corridor* (Boddy Lovering and Bassett, 1986) provides a useful historical analysis of economic change, employment and public policy in the Bristol sub-region.* Subsequent studies of economic restructuring, public policy and changes in governance practices include Stewart (1994; 1996a, 1996b), Oatley and Lambert (1995; 1998), DiGaetano and Klemanski (1993) and DiGaetano (1996).

---

* Bristol is also one of the four integrated city studies to be carried out under the ESRC's Cities: Competition and Cohesion Research Programme started in April 1998.

9

Rescripting urban regeneration, the Mancunian way

IAIN DEAS, JAMIE PECK, ADAM TICKELL, KEVIN WARD
AND MICHAEL BRADFORD

Introduction

When Central Manchester Development Corporation (CMDC) closed its
doors in March 1996 it did so proclaiming 'eight years of achievement'
(CMDC, 1996). The hyperbole was not entirely misplaced, for in the period
since CMDC's establishment in 1988 the southern edge of Manchester's city
centre has been thoroughly transformed in physical, economic and cultural
terms. But just as CMDC left its mark on the material landscape of
Manchester, it also played a part in an equally far-reaching transformation of
the city's political landscape. CMDC contributed to the normalisation of an
entrepreneurial mode of urban governance in Manchester, specifically playing
a central role in the City Council's embrace of more pragmatic and pro-
business styles of working. Once viewed as a bastion of municipal socialism,
Manchester City Council underwent a pronounced political realignment in the
wake of the 1987 General Election, shedding a tradition for oppositional and
minoritarian politics, and cloaking itself instead in a new proto-Blairite
approach based on 'can do' self-assurance, realism and flexibility.

This chapter argues that the imposition of a UDC was a pivotal moment in
the emergence of a new entrepreneurial politics in Manchester. We argue that
CMDC played an important role in the 're-corporatisation' of local governance in Manchester, a city which has arguably experienced a more fundamental redrawing of institutional relationships and structures than most over
recent years. This shift from welfarist concerns of 'defending jobs, improving
services' (in the words of the old City Council slogan) to the entrepreneurial
credo of 'making it happen' (as the new slogan has it) has been comprehen-
sively catalogued (Peck and Tickell, 1995; Quilley, 1995; Cochrane *et al.*,
1996; Tickell and Peck, 1996; Ward, 1997b, 1998). So too has the bombard-
ment of the city with a variety of urban regeneration initiatives, many of
which themselves echo the transition from (social) distributional to
(economic) development concerns (Kitchen, 1997; Lovatt, 1996; Randall,

1995; Tye and Williams, 1994; Williams, 1995, 1996). Within this broad context, we attempt here to trace the distinctive impacts of CMDC, both in the more direct sense of material and physical change and in terms of the wider regeneration agenda in the city. First, we examine the concrete achievements of CMDC. Second, we explore the role played by CMDC in transforming the city's politico-institutional landscape. The chapter is concluded with an assessment of the overall impact of CMDC on regeneration politics, Manchester-style. Throughout as necessary, we use extracts from interviews conducted as part of a number of projects investigating urban policy in Manchester (see acknowledgements below)

Incorporating development: CMDC and the 'property solution'

Central Manchester Development Corporation was established in June 1988 as one of three UDCs created in the wake of the Conservative government's re-election in 1987. Mrs Thatcher's election-night commitment to 'do something' for disadvantaged urban areas saw the UDCs' property-led method of regeneration extended to one of the strongholds of local Labourism, Manchester, alongside new initiatives in Bristol and Leeds. In contrast to earlier waves of UDCs, economic conditions in the three mini-UDCs were relatively buoyant as urban Britain began tentatively to recover from recession and started to benefit from the Lawson boom of the mid- to late-1980s. In Manchester – and, in particular, city centre Manchester – this sometimes hesitant upswing was based around growth in the financial and producer services, and a retail sector reinvigorated by credit-fuelled consumer spending (see Peck and Emmerich, 1992). However, there remained some significant blocks on development, with the southern fringe of the city centre, in particular, appearing immune to the more general resurgence of the city's economy. In common with other UDCs, this was attributed to a combination of contumacious physical obstacles to development and what the Conservatives in central government saw as an inability on the part of existing public sector institutions to work with the private sector. As Manchester began to emerge from the shock of the haemorrhage of traditional manufacturing jobs in the early 1980s – and as economic development policy increasingly sought to build upon the partial and uncertain gains in service employment as a means of revitalising the city's economy – removing these remaining constraints to development became a priority.

CMDC was the smallest of the UDCs, charged with the regeneration of an area of 187 hectares that skirted the southern fringe of Manchester city centre but which was also hemmed in by some of the city's most impoverished areas (see figure 9.1). To the east lay a corridor of pronounced economic and physical decline running through Openshaw and Beswick, locations of serial policy intervention, from the Urban Programme-funded initiatives in the late

1970s to expenditure in the 1990s in support of the city's 'sports-led' regeneration. The southern and western fringes of the UDA, meanwhile, were encircled by a string of equally impoverished areas, from Hulme and Moss Side running west to Salford. Significantly, though, the boundaries of CMDC were drawn to exclude these areas, in a conscious effort to ensure that the UDC remained focused and single-minded in its efforts to secure property-led

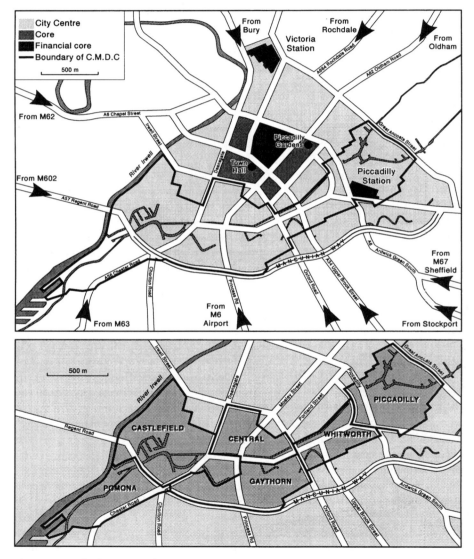

Figure 9.1 *The location of CMDC*
Source: Robson *et al.*, 1998a

redevelopment and, at the same time, avoiding the supposed distractions of dealing directly with the residents of impoverished communities. Instead, the designated area was drawn deliberately to embrace a swathe of fringe city centre land, much of which had proved resistant to the more general revival of the city centre economy.

The UDA was delimited to incorporate a relatively narrow corridor of land. The area around the former Pomona docks – fringing the major regeneration areas of Trafford Park Development Corporation and Salford Quays beyond the city of Manchester's boundaries – formed the western extremity of the UDA. To the immediate east lay the Castlefield area, home to a web of canals and railways, the historic Roman centre of the city and an area for which tentative regeneration plans had already been laid by the City Council and the former Greater Manchester Council. On the eastern side of the north-south Deansgate axis, which bisects the city centre, the designated area continued along the Whitworth Street corridor to the south of the city's retail core and to the north of the higher education precinct and the Mancunian Way motorway. The under-developed area around Piccadilly station on the eastern edge of the city centre, fringed by the city's inner relief road, formed the eastern boundary of the designated area.

The designated area was seen as one sufficiently large to accommodate the demand for development, which, it was hoped, would radiate outwards from the city centre. At the same time, it was also seen as sufficiently modest in scale to allow resources effectively to be concentrated in order to trigger development and to ensure that developer interest would never entirely be sated. As a result, the northern periphery of the city centre along the inner relief road linking Piccadilly and Victoria stations was consciously excluded from the UDA (Robson *et al.*, 1998a). In hindsight, this was a decision which, in light of the subsequent recession in local land and property markets, was later to prove crucial in maintaining a flow of capital receipts, and in sustaining a higher degree of developer confidence than existed in many other UDCs at that time.

Although over half of the UDA was classed as 'developed land', at its inception CMDC was faced with some sizeable physical challenges, not least of which was the one-quarter of its area defined by consultants as 'derelict, 'disused' or 'underused' (ECOTEC, 1988). Much of this took the form of derelict tracts of land, a substantial proportion of which suffered from problems of contamination associated with earlier uses. Investment was needed to assemble and prepare developable sites, and to bolster land supply in an area which potentially might benefit from emerging development pressures elsewhere in central Manchester in the late 1980s (Robson *et al.*, 1998a; ECOTEC, 1988). At the same time, these intractable physical obstacles to development were further complicated by a highly fragmented pattern of land ownership, which served further to undermine the area's development prospects.

CMDC's area differed in character from other UDCs in a number of important respects. Significantly, its establishment was premised on an assumption that the designated area was already home to some significant assets around which regeneration activities might be developed. The extensive networks of (frequently derelict) canals which traversed the area were seen as one such asset upon which the area's revival might be built. The remnants of Roman ruins in Castlefield, likewise, were regarded as a potential base for new tourist attractions. And some pioneering policy interventions had already yielded benefits in the way of some of the first residential developments to be completed in the city centre. Together, these constituted significant resources, as one CMDC official conceded:

> We have achieved a lot more than was first thought . . . Now why did we do that? Not because we are the most brilliant, all-singing all-dancing development corporation in the world, but because we had some pretty good raw material to work with . . . We had a rather telling mixture which characterised most of the successful regeneration initiatives in the '80s and the '90s: water; cleared site; large, ornate and sometimes attractive listed buildings.

While the establishment of earlier UDCs also presupposed the existence of a range of development opportunities, CMDC differed from its predecessors in that the relative strength of much of the service-based city centre economy already constituted a major asset on which it might draw. Indeed, before CMDC's designation office rental levels were already healthy, with 'best rents' having increased by 80% between 1984 and the establishment of the UDC in 1988 (Manchester City Council, 1995). Reflecting this underlying property-market buoyancy, central government's initial intention was that CMDC (in common with its siblings at Bristol and Leeds) would be granted more modest resources than the first two generations of UDCs. Central Manchester, Leeds and Bristol, according to Nicholas Ridley, the then Secretary of the State for the Environment, were to be 'cashless UDCs', where initial investments in unblocking land supply and in bolstering developer confidence ultimately would be offset by receipts flowing from land and property sales.

As a result of this putative stock of assets, and the relative well-being of the city centre economy more generally, the initial expectation was that CMDC would be granted comparatively miserly resources. Over its eight-year existence, CMDC's grant-in-aid from central government amounted to £82.1m, augmented by an additional £5.1m from European Regional Development Fund monies. These provided the bases for total expenditure of £100.6m between CMDC's establishment in 1988 and closure in 1996 – a figure in excess of the £71m in the Leeds mini-UDC, but less than the £112m in Bristol (Robson *et al.*, 1998a) see figure 9.2.

In common with other UDCs, CMDC was conceived as a means of administering, in the language of the time, a 'short, sharp shock' to reawaken lethargic local property markets. More prosaically, its formal objectives, based on those of the 1980 Local Government Planning and Land Act, were four-fold: to bring back into use land and property; to support new developments sym-

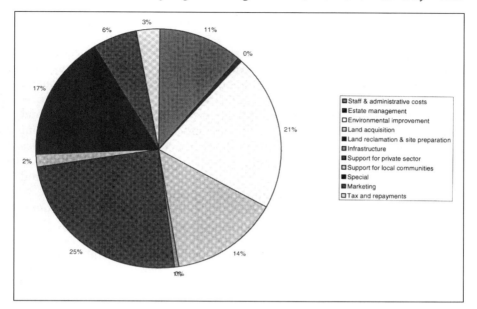

Figure 9.2 *CMDC spending by category, 1988/89 to 1995/96 (%)*
Source: CMDC Annual Reports

pathetic to incumbent buildings; to utilise private finance to underpin (re)development; and to improve the environment of the UDA. To these were added two rather more strategic, and locally-sensitive, objectives. First, the geographical extent of the city centre was to be increased through the stimulation of development around its southern and eastern fringe by releasing developable land and buildings, by pump-priming major property development and, less tangibly, by nurturing a more propitious climate in which developers could operate. In adopting this physical 'bricks and mortar' approach, one of CMDC's most critical objectives was to support the incipient growth of the city's financial and producer services sectors, which were increasingly seen as critical to the maintenance of Manchester's rather hesitant economic revival in the late 1980s (see Peck and Emmerich, 1992).

However, the objective of supporting the city's financial services sector was complicated by the lack of an adequate supply of accommodation of the required scale or quality, and recent additions to the city's stock of office space largely comprised refurbishment rather than construction (ECOTEC, 1988; CMDC, 1990; Manchester City Council, 1990; PA Cambridge Economic Consultants, 1991). At the same time, much of the city's financial services sector was shoehorned into the narrow 'square half-mile' financial district, with limited opportunities to develop in more peripheral city centre locations (Robson *et al.*, 1998a). At its establishment, CMDC was seen as crucial to boosting the supply of office space by removing some of the impediments to development within its designated area. This, in turn, would allow the office-based service sector to continue to expand, in both geographical and employment terms.

A second strategic objective sought fundamentally to alter the functional characteristics of the southern edge of the city centre, principally by supporting a series of housing and leisure-based developments. This, it was argued, would contribute towards the creation of a more diverse city centre, with new uses complementing the established retail sector and the revitalised office and commercial property markets generally, but – crucially – also positioning and branding much of CMDC's territory as a distinctive area within the broader city centre (CMDC, 1990). Significantly, CMDC's strategic emphasis on leisure and housing developments on the one hand, and financial and producer services on the other, would prefigure in a variety of ways the City Council's emerging development strategy, itself strongly attuned to the stimulation of professionalised services and cultural innovation (see Quilley, 1995; Peck and Tickell, 1995; Mellor, 1997).

In seeking to realise its objectives, both through individual and joint activities, CMDC would also play a part in establishing a new modus operandi for regeneration in Manchester. Entrepreneurial 'wheeling and dealing' in local land and property markets formed the core of CMDC's strategy. Operating with an overtly commercial bent – which was held to contrast markedly with the traditional approaches adopted in public-sector regeneration strategies, particularly those previously dominant in Manchester – the purchase and sale of land and buildings was intended rapidly to stimulate a variety of high profile property developments, which in turn would promote the area's image in the eyes of developers and accelerate the overall pace of regeneration. The aim was to ensure that property developers and other agencies were made aware of initial successes such as the Castlefield area and the Whitworth corridor, where opening negotiations with developers had already taken place, some modest land assembly had occurred and, in the case of Whitworth Street, some residential development had been completed, much of it under the auspices of the Phoenix Initiative and Greater Manchester Council. CMDC was able quickly to build upon these initial gains, allowing it rapidly to produce some highly visible outputs and, in doing so, to establish its credentials as a 'big hitter' in the local property market, and in the local political arena. Such tangible, and quite often *literally* concrete, outcomes were to become the main currency in Manchester's realigned local policy regime (Peck and Tickell, 1995).

Initially at least, there is little doubt that the injection of substantial resources in the wake of CMDC's establishment gave a considerable fillip to local land and property markets. In the first two years of CMDC's existence, as Robson *et al.* (1998a) show, rental values in the UDA accelerated past those in the rest of the city centre and continued to out-perform the wider area at the peak of the early 1990s property boom (see Figure 9.3). Much of this initial spurt may plausibly be attributed to increased developer optimism in the immediate wake of CMDC's establishment – and, thereafter, to the health of the wider regional property market – but it is also clear that the development climate in CMDC's area did warm initially, and did so more intensely and speedily than did the rest of the city centre. Moreover, CMDC can justifiably point to the development of high-quality office space, for example at the Great Bridgewater Hall, as testimony to their attempt not merely to engage in another round of speculative building, but strategically to counter the structural shortage of prestige premises which (arguably) existed throughout the city centre in the late 1980s. Developments such as this – and others such as the new Grand Island office development for the British Council at Gaythorn – went some way towards satisfying what consultants had highlighted as the overall paucity of new (as opposed to refurbished) office space coming onstream in the city centre (PA Cambridge Economic Consultants, 1991).

In fact, CMDC's forays into the local property market yielded some striking gains in the way of individual office developments. The near £27 million of private office investment levered from £7m CMDC money stands comparison with other UDCs and other regeneration agencies. At the same time, the £100m of investment in 70,000 sq. m. of office space for which CMDC provided no direct assistance might also be viewed as an indication of the degree to which the activities of CMDC boosted the property market in a way which might just prove durable in the absence of the supporting prop of

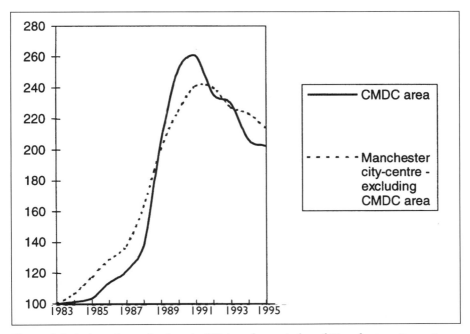

Figure 9.3 *Index of rental values in UDA and remainder of Manchester city centre, 1984–95*
Source: Robson et al., 1998a: Investment Property Databank

central government resources. A less optimistic assessment is also possible. After initial successes, the scale of the downturn in the area's property market – which outstripped the city centre as a whole – suggests that improvements were precarious, and that the property market remained more fragile than in the rest of the city centre. This is borne out by some spectacular individual reverses, the most noteworthy of which came with the protracted (and, at the demise of CMDC, still unresolved) wranglings over the future of the pivotal Great Northern Warehouse development, a proposed £100m office and festival-shopping complex – later also to include housing and a hotel – based around a historic (and listed) warehouse. Complex and ambitious proposals inherited from the former Greater Manchester Council were altered and scaled-down by CMDC, but the project repeatedly stalled as development and political climates waxed and waned, as difficulties ensued with the use of compulsory purchase powers, and, latterly, as the specially-formed Civic Society proved an additional, irritating check on developer ambitions. Another embarrassment was the Joshua Hoyle building in Piccadilly, again a historic, landmark building which CMDC also struggled to redevelop. Purchased amidst the usual hyperbole near the peak of the city's property market upswing in 1988, the building was saved from imminent demolition and was to form the cornerstone of CMDC's strategy for the Piccadilly sub-area. While this very act of purchase was intended partly as a signal of CMDC's muscle in the property market, the subsequent downturn in the market, and resultant lack of developer interest, left CMDC with a prominent eyesore on its hands. Rather than reaffirm CMDC's credentials amongst property developers, the inability to trigger the redevelopment of the building came to be a major embarrassment to the Corporation, serving graphically to illustrate the impotence of even a well-resourced body in the face of a sluggish property market. Had it not been for the safety net provided by post-windup funds from English Partnerships, the eventual agreement to redevelop the site as a hotel might not have occurred and CMDC's legacy in this respect may have been even less salutary.

Such reverses are common to most UDCs, and it is difficult to see how they could be eliminated entirely, particularly in the context of the less than propitious economic circumstances in which CMDC operated during the latter half of its existence. And for every such misfortune, CMDC can also lay claim to a number of prominent achievements. Nonetheless, that CMDC's property-led strategy proved less than entirely successful, despite the considerable funds at its disposal, is amply illustrated by property market data which show that, after initial gains, rental values in the UDA plummeted by almost one-tenth, at a time when values in the rest of the city centre were broadly stable. This might suggest an area which is more sensitive to shifts in wider property market conditions, but it also implies that, despite CMDC's early efforts, the area remained insufficiently robust to weather the cooling of broader property market conditions.

That rental levels in the UDA remained disproportionately capricious undermines the claim that CMDC effected a sustainable transformation in property market conditions. At the same time, there is also a sense in which CMDC may have cast a development shadow over adjacent, but non-designated parts of the city centre. Towards the end of CMDC's life, there was already evidence to suggest a surplus of some 1.3 million sq. ft. of office space in the city centre, some portion of which could be attributed to the national property market recession, but which at least one study concluded was the result of the displacement of firms from core to peripheral city centre sites, the latter of which included the UDA (Law and Dundon-Smith, 1995). By 1994 and 1995, moreover, take-up of newly constructed office space – almost all of it on the edge of the city centre, and much of it in CMDC's area (Manchester City Council, 1995) – was said to be substantially higher than that for existing office space. Despite the partial property market recovery, and despite CMDC's initial protestations to the contrary (CMDC, 1990), there is considerable evidence to suggest that UDC-led activities were accentuating a glut of office space in the city centre (Grimley, 1995; Richard Ellis, 1996).

Migratory flows from core to peripheral city centre sites included both small firms for whom cheaper rental levels were a powerful enticement, but also larger financial and legal firms attracted by the promise of purpose-built accommodation and, significantly, the lure of an increasingly prestigious address. However, criticisms of the volume of displacement engendered by CMDC's activities have limited weight, given that the Corporation's fundamental aim was to effect precisely these sorts of moves, thereby enabling the city's financial services sector to retain, and increase, its competitiveness. Despite signs of a southward shift in the commercial centre of gravity of the city, there are equally strong counter arguments that new developments in the UDA gave the office sector a greater degree of coherence and, in doing so, may have served to bolster the overall competitiveness of the city. Moreover, this is supported by research which suggests little negative impact on surrounding areas as a result of the displacement of firms. Amongst companies new to the UDA during CMDC's lifetime, Robson *et al.* (1998b) found in a study of property market vacancy chains that only 22% comprised relocations which resulted in vacancies at the originating premises. By contrast, most 'new' firms constituted net additions to the local economy, comprising either new starts (29%), off-shoots of existing companies located elsewhere (44%), or relocating firms which had been replaced at their location of origin (5%). While office-based employment was predominant in the displaced firms – giving further credence to criticisms about CMDC's impact on the existing office core – these are figures which compare favourably with UDCs in Leeds and Bristol, where 30% and 27% of 'new' companies in the respective UDAs were the result of displacements from non-designated areas (Robson *et al.*, 1998b).

Overall, then, CMDC met with mixed results in its efforts to effect a modest revival in property sector fortunes. Viewed within the context of what, for Manchester in the 1980s and early 1990s, was the unprecedented sum of public money at its disposal, this raises legitimate doubts about the extent to which CMDC met the 'value for money' criterion on which the governments of the time were insistent that regeneration agencies should be judged. Nevertheless, in terms of property market performance, CMDC could point, with some justification, to a final balance sheet which betrays some unsullied accomplishments. Private investment levered – in the eyes of UDCs and government, the principal measure of success – amounted to £303m, exceeding the initial expectation of consultants (ECOTEC, 1988). Although data on commercial floorspace and land reclamation offer a less positive picture (Table 9.1), this has to be viewed in the context of the unhelpful property market within which CMDC worked through much of its existence. And while criticism of the reliability of these crude output data is difficult to refute (Shaw, 1995 also see Chapter One), our earlier analysis of the limited amount of quantitative data with which it is possible to measure impact suggest that – notwithstanding isolated difficulties over particular developments and the continuing fragility of the local economy – CMDC did manage to improve property market performance.

Judged against the task set for it by central government, it is likely that CMDC will be seen as one of the more successful UDCs. Yet this largely positive assessment fails to capture the broader regeneration picture. In contrast to earlier regeneration initiatives in Manchester, CMDC saw its remit as one centred largely on the objective of securing a revival of property market fortunes. To a limited extent, this simply reflected the thrust of national regeneration policy in the late 1980s. But this ignores the experience of some other UDCs, such as that in Tyne and Wear, which pursued a more generously defined form of regeneration, and one which paid more than just lip service to broader social or redistributional goals (Deas and Robinson, 1994).

The impact of CMDC's housing regeneration efforts reflects this stance. CMDC helped to stimulate an appreciable upswing in what had previously been an inert housing market. Indeed, housing development in the city centre as a whole had been restricted to sporadic developments such as those at St John's Gardens and atop the Arndale shopping centre, and initial efforts on new-build at Piccadilly Village and refurbishments along Whitworth Street. Total population in the UDA at CMDC's establishment was estimated to amount to no more than 250 people (CMDC, no date). Moreover, there had been no concerted effort to utilise public funds to boost this total, in marked contrast to neighbouring Salford, where substantial resources had been injected throughout the 1980s in an effort to refurbish local authority-owned stock and promote owner-occupation (Bradford and Steward, 1988).

CMDC direct expenditure of £13m (excluding monies spent on related environmental improvements) helped generate in excess of 2,500 dwellings in the UDA, and in that sense alone drastically altered the character of much of its area. As Robson *et al.* (1998a) show, the proportion of land-use in the UDA claimed by housing increased five-fold between 1989 and 1995, an achievement which is all the more striking when calibrated against the rest of the city centre, where residential acreage dwindled. Indeed, the contraction in

Table 9.1 Land and property development targets and outputs

	ECOTEC target (1988/89–1993/94)	Actual (1988/89–1993/94)	Final (1988/89–1995/96)
Land brought back into productive use	31.1 ha	29.4 ha	35.0 ha
Commercial floorspace supported	193,600 sq. m.	128,400 sq. m.	138,600 sq. m.
Canals and waterways improved	8 km	9.4 km	13 km
Housing units created	471	1270	2583

Source: ECOTEC, 1988; CMDC, 1996.

residential land-use in the non-CMDC portion of the city centre – at a time when the City Council was keen to 'talk up' the housing market – was such that by 1995 it lagged behind the total area of housing land in CMDC's area. However, the most compelling illustration of CMDC's success in supporting the housing market is provided by the UDA's experience in the wake of de-designation. The continuing buoyancy of much of CMDC's area evidences the durability of these initial improvements, with non-subsidised housing in Castlefield and in Gaythorn/Whitworth Street – the first of such housing in the city centre – providing graphic testimony to CMDC's success in this respect. City centre housing is now a high-profile element of the Council's regeneration strategy. The local authority is anticipating, and indeed is now planning for, a substantial increase in the city-centre residential population in coming years.

Yet questions remain about the extent to which these achievements complement the putative distributional goals which, latterly, UDCs were encouraged to embrace. CMDC's housing objectives did make reference to the goals of tenurial balance and affordability for low income groups (CMDC, 1990), but the majority of new housing constructed during its lifetime, and subsequent to its demise, overwhelmingly comprised owner occupied stock marketed at affluent young professionals. Such development was important in its own right, helping to create a population – overwhelmingly comprising affluent, childless households, many working in the city centre, and many with second homes (Robson *et al.*, 1998a) – with which to underpin the development of a related infrastructure of city-centre services. Housing development at Castlefield, for example, was essential in anchoring a related range of up-market bars and in bringing a more general sense of vibrancy to the area. Alongside this, the emphasis accorded to social housing was barely discernible, despite CMDC's initial plans. Leaving aside the isolated social housing development at India House, low income groups were conspicuously absent from the UDA, and even the socially more mixed development at Piccadilly Village was underway before CMDC's establishment.

The way in which the housing market evolved offers a stark illustration of the success with which CMDC met its tightly delimited responsibilities. In terms of employment creation, too, the question is not so much whether CMDC's satisfied its objectives, but whether its brief was defined too restrictively. While net employment creation – amounting to almost 5,000 extra jobs – failed to match initial forecasts, much of this might be explained by the national recession, and CMDC can claim to have boosted job numbers in all but one of the years of its existence (the exception being 1991/92). But as with its 'achievements' in triggering housing development, questions remain about the equity of job creation. CMDC officials appear to have made a conscious decision not to engage with any of the impoverished areas which girdled the UDA. The impact of this strategy is clear from the result of one survey which found the proportion of employees drawn from deprived areas – at just under one-fifth of total employment – to be broadly similar amongst firms which pre- and post-dated the establishment of CMDC (Robson *et al.*, 1998a). In part, this reflects the predominance of skilled and white collar jobs amongst new employment generated. But it is also the results of the decision not to intervene in the labour market, preferring instead to view this as the exclusive and separate province of the Training and Enterprise Council. Once more, this decision could be justified with reference to the narrowness of the enabling legislation, but again the more interventionist stance of the likes of Tyne and Wear Development Corporation stands as a useful point of contrast (Deas and Robinson, 1994). The more general conclusion to be drawn, however, is that the impact of CMDC on the labour market (like that on the housing market) is emblematic of the more general outcome of its activities. While, for the most part, it met its core remit with conspicuous effect, its success is much more questionable when judged against broader regeneration criteria.

Developing incorporation: CMDC and the reconstruction of Manchester's regeneration agenda

If UDCs were intended ostensibly as a mechanism through which physical infrastructure might be regenerated and economic revival triggered, they also had a central role in the wider Thatcherite political project which increasingly sought to cast Labour-controlled municipal authorities as the principal impediment to the modernisation of urban Britain (see Imrie and Thomas, 1993b and c; Peck, 1995). Although largely based on electoral enmity and the legacy of decades of fraught central-local relations, the Conservatives feared that local authorities might derail the programme of fundamental reform on which they had embarked. In this respect, Manchester was viewed with deep and enduring suspicion. In common with Liverpool, Sheffield and, most strikingly, the Greater London Council, the city had assumed a rhetorical opposition to national Conservatism and had been vilified both nationally and locally for its high profile campaigns on behalf of minority communities. Yet despite the intense national disapprobation directed at the city for much of the first half of the 1980s, local electoral support for Labour remained firm: the apparent national electoral hegemony of the Conservatives was not matched in Manchester (see Edgell and Duke, 1991; Taylor *et al.*, 1996). Indeed, by the time of CMDC's establishment in 1988, with the combined strength of Liberals and Conservatives on the City Council amounting to less than 10% of all elected members, Labour dominance of local electoral politics was so pronounced that a group of dissident Labour councillors had come to constitute the *de facto* opposition.

There was, then, a clearly articulated political rationale for the establishment of CMDC. Despite a battery of measures to dilute the powers and responsibilities of local government, and concerted efforts by the Conservatives to disperse responsibility for the various aspects of local governance amongst a network of specialist non-elected appointed (and sometimes short-life) public bodies, Manchester remained a city in which electoral politics were dominated by the Labour Party, and in which policy-making was thoroughly infused with the key tenets of municipal Labourism. Alongside the objective of reviving the local property and land market, the establishment of CMDC was also premised on an expectation that it would attempt to dislodge deeply-ingrained attitudes towards regeneration and open up the City Council both to the influence of local business actors and to their means of doing business (see Peck and Tickell, 1995).

In view of this, the imposition of a UDC might have been expected to create considerable friction between a Labour-controlled urban local authority and Conservative-authored UDC. Indeed, there is a welter of evidence, from London Docklands to Teesside to Bristol, to support this fear (Brownill, 1993; Robinson *et al.*, 1993; Oatley, 1993). In Manchester, however, relations were much less fraught than might have been expected given the city's recent history of confrontational politics. By 1988, the city was already witnessing a conspicuous move away from the rough mix of traditional municipal Labourism and the radicalism of the more interventionist 'new urban left' (Gyford, 1985; Quilley, 1995), both of which dominated the agenda of Labour-controlled local authorities at a time when Thatcherism held sway nationally. In Manchester, this shift was prompted partly by the national climate of 'new realism' in the wake of Labour's defeat nationally in the 1987 election (Beynon *et al.*, 1993). Compared to the UDCs at London Docklands and Merseyside – both formed in 1981, at the height of Thatcherite aggression towards local authorities – CMDC's establishment was favourably timed. By 1988, Manchester's politics were on the cusp of dramatic change (Quilley, 1995; Cochrane *et al.*, 1996; Ward, 1997b). During the course of CMDC's existence, Manchester City Council, in common with other urban Labour local authorities, became increasingly at ease with the notion of utilising private finance to fund public-sector capital development projects in fields like health and transport; it became more sanguine about contracting-out regulated public services to private-sector providers; it came consciously to involve the private sector in local economic regeneration projects; and it became more sensitised to the collective 'business voice' (Peck and Tickell, 1995; Randall, 1995; Mellor, 1997).

At its establishment, however, CMDC was faced with a very different context: one in which it had quickly to establish itself as an affluent, upstart creature of central government, lacking the electoral legitimacy of the local authority. Mirroring the pattern in other UDC cities, Manchester City Council took the decision formally to oppose the 'imposition' of CMDC. The City Council objected on a whole range of grounds: the assumption of wide-ranging planning and development powers (some of them from the City Council); the lack of any local electoral mandate; the relative wealth bestowed on CMDC at a time of local authority retrenchment; and the emphasis accorded to property development at the expense of more overtly redistributional goals. In the eyes of the leading local politicians, CMDC was a body dominated by business people who had long been among the most vociferous critics of the City Council; it was a body whose membership contained one-third of the Council's Conservative group, but barely one-in-a-hundred of the Labour group.

In view of CMDC's strategy of demonstrating its financial muscle and commercial acumen in land and property markets at the earliest possible stage, it was imperative that it remove (or at least reduce) the obstacle provided by City Council hostility as quickly as possible. To do so, it might have followed the path of other UDCs and adopted a strategy of macho disregard and defiance of what some certainly saw as the City Council's outmoded, anti-commercial bent. That it opted to do otherwise, recognising the importance of the City Council, and attempting to mollify the ruling Labour group, was crucial in helping it avoid the marginalised existence which befell some of its less successful counterparts. Indeed, within a relatively short period of time the relationship between the two bodies had warmed sufficiently for the Leader of Manchester City Council, Graham Stringer, to declare that Central Manchester Development Corporation was now 'a quango . . . in danger of giving quangos a good name'. And this was to be a mutual volte face, as CMDC made similarly laudatory noises about the City Council:

> We have not worked alone. The roots of our success lie in the effective partnerships we have forged . . . [and] . . . central to our overall level of achievement have been the relationships at all levels with Manchester City Council. (Central Manchester Development Corporation, 1996, p. 5).

In part, this dramatic turnabout reflected the increasingly frenzied attempts to cultivate partnership arrangements at the end of the 1980s and the beginning of the 1990s (Peck and Tickell, 1994; Tickell *et al.*, 1995). Cross-agency collaboration offered agencies such as CMDC a strategic position that was far in excess of what might have been expected for a short-life body with resources which, though substantial, were modest in comparison with those at the disposal of the City Council. What is more significant, however, is the scale of this transition, and the speed at which it proceeded. Of the three cities with mini-UDCs, only in Manchester was initial antipathy transformed into willing and amicable co-operation (Robson *et al.*, 1998a). There are a number of reasons for this. Central to the explanation of the relative ease with which CMDC was accepted by and into the wider network of policymakers was the timing of its establishment, which coincided with the campaign to secure national support for the proposed bid for the 1996 Olympics. This allowed CMDC's usefulness, in the eyes of the wider coterie of policy influentials, to be established at an early stage: the deftness with which CMDC manoeuvred to construct a credible partnership, and the resources at its disposal, offered a compelling illustration of its value to the city as a whole. During the meeting at which it was decided that Manchester should go forward as Britain's Olympic candidate, it was claimed that the city's UDCs (Trafford Park and CMDC) could lever £2 billion of investment during the run-up to the 1996 Games (*The Times*, 20 May 1988), while the British Olympic Association (BOA) was deeply impressed by Bob Scott's vision of a Los Angeles-style 'private enterprise Games' (Hill, 1996), central to which, inevitably, would be CMDC. Manchester, it seemed, had been 'awarded' a UDC at a propitious moment, a point acknowledged by a senior CMDC officer:

> We . . . had the benefit of being in Manchester at the right time. [CMDC was established] at a time when a few of the key . . . 'movers and shakers' were around . . . – rather than the 'loose change merchants' – and . . . the city had just begun to develop a confidence again, having had a quiet period. Maybe prompted by the Olympic bid, maybe prompted by a variety of things, . . . but things had started to happen in this city in a way they did not happen before.

Relations between CMDC and the City Council were further cemented by a number of significant deals sealed in the early stages of its existence. Negotiations over the composition of the CMDC board and the choice of chairperson were to prove particularly significant in shaping the ease and speed with which CMDC was accepted into the array of networks emerging in Manchester in the late 1980s. Given the history of mutual mistrust between public and private sectors in the city, the choice of chairperson required delicate handling: the appointee had to be politically inactive (but with sensitive political antennae) and needed to possess the degree of commercial acumen necessary to reassure the business community. The perceived success of this appointment made CMDC's early dealings with Manchester City Council less awkward. Indeed, a senior member of the executive attributed the relative ease with which initial dealings proceeded between CMDC and the City Council to the individual characteristics of the chairperson:

> It was quite an inspired notion to appoint the particular chairman they appointed here because although his business roots and commercial roots were in Trafford Park, he was known on the Manchester 'scene'. He was not, if you like, tainted by the Manchester 'scene' but was sufficiently well known and appreciated by people in Manchester . . . and crucially by the leadership of the City Council, with whom he had been involved with on the Manchester Science Park board.

The timing of CMDC's establishment, coupled with the adhesive effect of the Olympic bid on institutional relationships in Manchester, and the judicious choice of senior personnel, helped to propel CMDC with a momentum denied to UDCs established in less favourable circumstances in other areas. At the same time, this was reinforced by the development of successful working relationships at more junior levels. The decision to contract-back to the City Council statutory development control powers which had been assumed at CMDC's inception was one initial decision which helped to build trust and co-operation at officer level, reinforcing the intimacy of the partnership which was emerging at executive level. While UDCs which opted to retain development control powers in-house were repeatedly undermined by conflicts with local authorities over planning issues of strategic importance (Robson *et al.*, 1998a), CMDC was able in the majority of instances (excepting isolated skirmishes over contentious plans for the Great Northern Warehouse) to proceed with the City Council's acquiescence on planning matters. The fact that CMDC was able to quicken the process of determining planning applications – and to meet its target of making decisions within eight weeks (Robson *et al.*, 1998a) – is evidence of the effective working relationships it was able to construct with the City Council.

The rapidity with which cross-institutional relationships were cemented was an important factor in helping CMDC quickly to generate tangible outputs in the form of agreements over property and land deals. But in a wider sense, it also served fundamentally to alter the city's regeneration agenda: amongst the burgeoning network of agencies in Manchester in the late 1980s and early 1990s CMDC played a key role in developing a new consensus about the priorities for urban regeneration and the most effective means by which these could be fulfilled. This, as much as the perhaps more palpable 'bricks and mortar' outputs of CMDC, is its most significant impact. The reconstructed regeneration agenda straddled a number of different elements, the adoption of which, in many cases, was spurred by powerful (and often pivotal) promptings from CMDC. A clear illustration of the extent to which this agenda shifted during CMDC's lifetime is provided by attitudes towards the goal of bolstering industrial development in the city. At its inception, CMDC pledged to promote industrial development, notably in the Pomona/St Georges area in the western fringes of the UDA (ECOTEC, 1988). This was an area located in relatively close proximity to Trafford Park industrial estate which exhibited characteristics less obviously in line with those of the city centre. By wind-down in 1996, however, it was the area in which fewest resources had been expended (amounting to a mere £47,000 on environmental improvements) and, as a result, fewest positive outputs generated. Ultimately, CMDC's initial plans for the area floundered, partly as a result of the difficulties encountered in tackling what was undoubtedly a complex and declining infrastructure of railways and canals. By 1994, plans for the area had been shelved and, according to DETR's official evaluation, the area 'remains much as it was at the beginning of the period' (Robson *et al.*, 1998a).

What is remarkable about this change of tack is not that, by 1994, industrial development, in effect, had been expunged from CMDC's list of aims and objectives, but that such a profound shift in policy had been able to occur, at best, with little or no opposition from Manchester City Council, or, at worst, with its active collusion. In the space of less than five years, the regeneration agenda had been altered so fundamentally that reactions to the decision to eschew industrial development ranged from indifference to hearty endorsement. While much of this reflected the national trajectory of regeneration politics at the time, the fact that Manchester was comfortable with this new agenda at a relatively early stage is, in part, due to the emollient, consensus-seeking approach which CMDC adopted, and which contrasted starkly with many other UDCs.

CMDC's attempts to diversify the city centre economy, and to develop new types of economic activity therein, were another dimension to this fundamental shift in regeneration policy and, after some initial (and largely token) worries also gained the assent of local regeneration policymakers. In order to secure this diversification, CMDC sought to nurture small firms in emerging economic niches by offering relatively cheap, short-lease properties. This proved important in underpinning the nascent growth of an array of creative industries around multimedia, music and design in developments such as Ducie House in the Piccadilly sub-area (which was later to close), or, more successfully, the cluster of software authoring firms around the CMDC-supported (and on-going) development at Eastgate House in Castlefield. As the words of one CMDC official confirm, this approach went beyond the haphazard opportunism – attempting to attract development, regardless of type or form – for which some UDCs have been criticised. Instead, attempts to diversify the city centre economy, and to cultivate particular types of economic activity, were part of a much broader, and more strategic, game plan into which, ultimately, many of the city's institutions were to buy:

> The city centre before we came was basically nine to five shops and retail, with a sprinkling of theatres and clubs. There are [now] an enormous number of bars, clubs, theatres, outdoor arenas, residential development to go with the office and the retail stuff that is coming to our area and the whole thing builds you a bigger group. That is what we have been about. It is about joining and making the city centre big enough to be a credible city centre internationally, with all the functions that it needs and ingredients that it needs. But it is also about giving the confidence to the rest of the city centre to promote itself in a positive way . . . So actually it lifts up more than just our area. It lifts up the whole [city].

This new and beguiling vision of a diverse and vibrant city centre constituted what Lovatt (1996) has termed a 'pro-culture growth strategy', reflecting a gradual melding of the culture-economy strategies of CMDC and those of the City Council (Quilley, 1995). The emerging regeneration agenda was one based around a redefined city centre, the role of which was to satisfy demands for consumer services through nightclubs and bars, sports stadiums and concert halls, and through a series of events and festivals, including the annual gay

Figure 9.4 *Riverside entrance to Bridgewater Hall, achieved by redirecting the canal*

Mardi Gras and Castlefield festivals, both of which are held in the CMDC area.

> This is [now what] Manchester is competing for . . . for seminars, conferences, it is competing for sports event; it is competing for the sort of thing that in the late twentieth century and beginning of the twenty first century form a major industrial city . . . Yes, there are still the major industrial activities, manufacturing activities which are important to attract to cities, but tertiary activity and the leisure, the business tourism, the conferences, the events . . . are the things that keep cities up there . . . You need the raw materials in your city to deliver that in the first place: . . . indoor arenas, national arenas, concert halls, convention centres, those sort of things, exhibition space, those have all, or all in the process of being delivered or delivered to Manchester within the last ten years and if you look forward to the next three or four years that process will provide a whole host of those sort of big infra-structure type things (Senior CMDC officer).

This pro-culture strategy was to prove seductive across much of Manchester's network of regeneration policymakers. The 20,000-seat Nynex events arena, for example, followed on from the CMDC-supported Bridgewater Hall concert venue. Likewise, proposals for residential development in refurbished mills and warehouses as part of the Great Ancoats Single Regeneration Budget

Figure 9.5 *Barça, Catalon Square. Co-owned by Mick Hucknall and illustrative of the transformation of the canal from a site of production to a space of consumption*

Challenge Fund initiative on the northern fringe of the city centre drew from CMDC's earlier experience in kick-starting the housing market in Castlefield and the Whitworth corridor. The City Council's strategic vision for the city centre also altered dramatically, and again bore the stamp of CMDC's influence. Alongside the incessant hype about creating a 24-hour city, the City Council embarked on a more corporeal attempt to mimic the success of the Castlefield area through a series of 'branded' cultural quarters traversing the city centre, from the Gay Village in the south of the city centre, to the cluster of challenging housing developments and nightlife in the Northern Quarter at the city centre's northern extremity.

In these ways, CMDC sought fundamentally to alter the agenda of the city's regeneration policymakers, both by subtly advancing its own interests in wider regeneration fora and by absorbing aspects of the increasingly entrepreneurial City Council strategy into its own development priorities. Equally, it also sought to alter the ways in which this agenda was to be fulfilled. In line with the tacit UDC objective of altering local political climates and challenging

Figure 9.6 *A new bridge across the Leeds-Liverpool canal, grant-aided by CMDC*

established patterns of institutional conflict, much of this was simply about confidence-building: attempting to foster better working relationships within and across private and public sector. According to one senior business leader, CMDC's strategy consisted of 'driving people into these relationships which are not stuffy, in which you create shorthand between two people . . . and they suddenly discover that they have more in common than they ever thought they did'. This process of 'creating shorthand' between élite actors implied a rejection of the stuffy formality of structured relationships in which affiliations such as public and private sector, or Labour and Conservative, were made explicit. But it also implied a critique of local democracy, public consultation and bureaucratic control, in which institutional formality was to be replaced by agency flexibility. Within this, the process of partnership working itself became markedly more fluid. In contrast to the phase of intra-organisational, bureaucratic decision-making, strategies were now formed through new channels of communication.

CMDC did not alone establish this new modus operandi but it was certainly one of the most prominent co-authors. Reflecting the CMDC approach, networking amongst élites in the public and private sector built up around 'shorthand' channels of communication and, in turn, helped facilitate rapid and 'business-like' decision-making across as well as within local agencies. As local political and business leaders explained, this more dynamic approach emerged out of a concern with:

> . . . the very slow process of the ideal form of democracy. It does take a long time. Short cuts have to be done in such a way that they're not too dictatorial. So, it's a form of . . . I can't say it's dictatorship, but it's *managing* democracy, I think, in a way which seems reasonably sensible to those who are trying to do it. To make sure that decisions get made and taken (City Challenge director).

> [The strength of the UDCs is] purely one of being able to operate without political interference . . . The problem with democracy is that . . . you get all sorts of stupid things that will happen. It's the decisions being made in council chambers with political pressures and not commercial pressures . . . The UDC is unfiltered by political influences . . . Take planning . . . At the end of the day the people making the decisions want to get back on the council . . . so how do they decide to do things? They decide to do things in a way which makes sure they get back on again next time (CMDC board member, private sector).

Manchester has built much of its contemporary reputation as a 'can do' city on these methods of élite networking, coupled with the new air of political pragmatism which now pervades most of its key local institutions. The city's numerous 'grant coalitions' have been conspicuously successful in their bids for public funds and (re)development projects (Cochrane *et al.*, 1996), in part due to their superior capacity to anticipate and reinterpret the priorities of external funding agencies (see Peck, 1995; Tickell *et al.*, 1995). Yet the strategy has not been without its costs. For all the bravado, Manchester's economic base remains weak (Peck and Tickell, 1997), while the city's new development agenda, which CMDC was instrumental in promoting, continues to subordinate social objectives to the overriding economic goals of property development. Problems of poverty and social exclusion continue to deepen in many parts of Manchester (see Griffiths, 1998), problems which were almost shunted off the policy agenda during the decade from 1987. It is only in the wake of CMDC's closure that these issues have begun to receive explicit attention.

Conclusion

CMDC can legitimately lay claim to numerous achievements in revitalising the city centre property market. It successfully triggered an on-going series of housing developments; it brought a greater sense of coherence to the city's commercial office market; it restored developer optimism in an area from which property interests had frequently recoiled; and it laid the foundations for some important *grands projets* which added a tangible degree of credence to the hype surrounding the city's revival more generally. In these respects, CMDC stands out as one of the more successful UDCs; to a large extent, it fulfilled the rather penurious remit set for it by Conservative governments.

At the same time, however, what is most striking about CMDC is the extent to which (in some senses) it went beyond this narrow, central government-imposed brief. At its outset, CMDC's implicit ambition was a bold one, which went some way beyond the more modest and non-strategic goals adopted by some other UDCs. While many of its counterparts in other areas chose not to develop longer-term 'visions' for their areas, CMDC cast for itself a central role in supporting a wide-ranging transformation of the city's economy from one based on declining manufacturing to one based on prospering services and, in particular, financial and producer services. In attempting this, CMDC met with a degree of success. The log-jam of office development which threatened to build up was successfully averted as many of the physical impediments to development were removed, allowing the financial services sector to continue along its hesitant pattern of growth, checked only by the continuing vagaries of the international economy.

A less tangible, but perhaps more enduring, aspect of CMDC's legacy has been to influence the trajectory of regeneration politics in the city. In this sense, too, CMDC attempted to go beyond its tightly-defined remit and, in the longer-term, to alter the outlook and culture of institutions in Manchester. In contrast to many of its counterparts – and, in particular, in contrast to the other mini-UDCs at Leeds and Bristol – CMDC developed a rather more expansive agenda than might have been expected from a strict translation of the duties set for it by central government. CMDC sought to develop a role as the leader of a pro-growth coalition which, initially, had coagulated around the city's Olympic bid. It saw its remit as extending beyond its formal geographical boundaries, while promoting itself in a transformative role with respect to local political and institutional relations. CMDC did not fit the crude UDC stereotype of a glorified estate agent concerned only with piecemeal, opportunistic property development. While other UDCs were concerned mainly to ensure development proceeded as quickly apace as was feasible in prevailing economic conditions, CMDC helped inculcate a longer view which saw the city centre diversifying from its traditional, narrow role as a centre of retailing and commerce (both of which would continue to be reinforced) towards a more broadly-based one which also encompassed prestige events, visitor attractions and the marketing of historical assets.

That CMDC was able to exert such a crucial influence reflected the timing of its launch. CMDC arrived at a time when the city's policy community was embarking – somewhat hesitantly – on the Damascene conversion from the welfare-driven agenda of traditional big-city Labourism, towards the pro-growth economic development ethos characteristic of most UK municipal government in the 1990s (see Stewart, 1994; Cochrane *et al.*, 1996; Ward, 1997b, 1997c). After a troubled start, CMDC's success in supporting a series of notable property-based projects and, more strategically, in effecting a broader repositioning of the role of the city centre, offered a useful exemplar with which to inform future strategies. In marked contrast to the rather more restricted roles developed by other UDCs, CMDC was able successfully to position itself as the lead agency in an incipient growth/grant coalition, around which other agencies and partnerships were able progressively to cluster over time. CMDC saw itself as central to the development of a new and different regeneration agenda, the potency of which would be underscored by the widespread support that the Corporation generated amongst a whole range of key regeneration 'players'. It is the boldness of this attempt to re-script the city's regeneration agenda – and to carve out a role for itself as lead player – which marks out CMDC amongst UDCs. In the absence of CMDC's catalytic role, it remains questionable whether the transformation of Manchester's local economic development agenda would have been able to proceed quite so dramatically, or quite so speedily. CMDC smoothed the transition, while being an integral ingredient in the change itself.

In other senses, however, the role which CMDC carved out was rather less expansive. While CMDC (again, unlike most other UDCs) did look beyond the geographical and functional boundaries set for it by government, this was largely confined to bridge-building with other institutions, trying to alter established ideas about regeneration and, ultimately, attempting to create a new model to which future regeneration strategies would work. However, this was a model premised on the desirability of property-led regeneration, and one which largely neglected the social consequences of property-led development. While the grand scale of CMDC's politicking marks it out as one of the more ambitious UDCs, the narrowness of the means by which regeneration was pursued shows it to be one of the least ambitious and most conventional of the UDCs. The inclusivity which it applied to partnership-building did not apply to its links with residents of neighbouring communities, the majority of whom remained marginal throughout CMDC's lifetime. While CMDC promoted itself as a pioneer of local neo-corporatist politics – a 'big tent' approach ten years ahead of its time – this excluded any attempt to engage with residents of adjoining impoverished communities. In this sense, CMDC lags behind some other UDCs (Tyne and Wear again being a notable example) which attempted to institute a whole series of mechanisms through which property development benefits could genuinely cascade to local residents. By contrast, CMDC opted to eschew any significant intervention to promote more equitable labour market outcomes; it decided not to involve residents in UDC decision-making; and it chose, in the main, not to assist with the development of any social infrastructure more generally.

A decade and more on from the inception of CMDC, like other UDCs it seems a rather curious entity, reflecting a unique national political and economic zeitgeist. The blinkered emphasis on growth – at the expense of any consideration of wider distributional effects – and the single-minded focus on securing property developments now seem historical oddities, increasingly out of step with emerging orthodoxies of social inclusion and the pursuit of more broadly-based goals for regeneration (see DETR, 1997). In the context of the 1980s, however, UDCs were about more than simply the regeneration of the physical and economic fabric of narrowly-defined parcels of land; they were also intended to transform institutional relationships and established outlooks on economic development. In both respects, CMDC can justifiably claim profoundly to have altered *the Mancunian Way*.

Acknowledgements

Kevin Ward acknowledges the support of the ESRC (award number R00429434350). Jamie Peck and Adam Tickell thank the Nuffield Foundation for their support for the research project 'The politics of business leadership in Manchester'. Adam Tickell thanks the ESRC for its support for the research fellowship 'Regulating finance: the political geography of financial services' (award number H52427001394). Finally, all five authors would like to thank the eighty or so individuals who were interviewed on the various projects. The usual disclaimers, of course, apply.

Further reading

Writings on Manchester's experiences are limited and the best overview of the CMDC is the 'Robson report' (1998a). ECOTEC (1988) provides a useful review of CMDC's initial rationale and strategies. Broader processes of change in Manchester's economy and politics are documented in Peck and Emmerich (1992), Peck and Tickell (1995), and Ward (1997a).

PART III

Prospects for Urban Policy

10

New deal or no deal for people-based regeneration?

BOB COLENUTT

Introduction

This chapter addresses the age old question; does politics change anything? Regeneration for people was not the main priority of the government during the 1980s and was credited with making inequalities in urban areas worse by supporting market-led solutions. Urban Development Corporations epitomised a top down, property driven approach and, to the relief of many, are being wound up, now replaced with new regimes for funding and managing urban development. But does the different political and economic context of the end of the millennium mean that there will now be a more bottom up, more needs based approach to regeneration?

There is no doubt that there is a new climate of social and to some degree cultural tolerance; and also a resurgence of regionalism and localism since the end of the Thatcher era. There is hope, there is a new political language of inclusion – but will this lead to more people power and a more inclusive mode of urban renewal?

The chapter explores both the opportunities and the barriers to people-based regeneration in this new era. It examines, in particular, the crucial role of one of the principal gatekeepers in regeneration – local authorities.

The term 'people-based' regeneration is used here with some reservation. Usage of the word 'people' by politicians and opinion formers appears to be moving ahead of the over-used 'community'. Until recently, 'people-based' regeneration could have been easily understood as renewal that was targeted at communities of disadvantage and need. Moreover, it implied solutions to urban problems generated 'from the base' by grass roots community, voluntary, trade union organisations and local authorities – rather than by market forces or agencies of central government, such as Urban Development Corporations or City Challenge Boards

Yet as we come to the end of the 1990s, an appeal to 'the people' is much more complicated. It does not necessarily imply a socially progressive movement, nor does it have social class, non-market or non-property implications.

But in spite of these complexities, class, race, social disadvantage and exclusion remain critical factors in any definition of regeneration and thus the label 'people based' is used here.

Urban policy – dateline May 1997

The 1997 General Election gave every appearance of a break with the familiar policies and practices of the previous two decades. A value shift seemed to take place, from materialism to a more caring society, from social exclusion to social inclusion, and from private accumulation to concern for the public realm (or at least a new mixture of the two). These shifts were expected to have a major impact on all areas of Government and indeed to the way in which both central and local government and agencies attached to them went about their work. Urban and environmental policy, regeneration programmes, and funding regimes would all be affected. At a grass roots level in urban areas in particular, it was expected that the centralising tendency of government, the imposition of property led renewal, and of top down planning, would be halted and put into reverse. It was not obvious what was coming next, but most practioners thought that at the very least relationships between local and central government would be easier, that there would be greater community participation, and that funding would be targeted at deprived areas (see for example, DETR, 1998a and b).

On closer inspection, demand for these changes was not universal. Quite to the contrary. At a strategic level, major public and private sector interests had developed a robust structure for networking and partnership which many of the key regeneration players were happy with. They were resistant to pressures for change that were coming from the grass roots.

By the mid 1990s, most of the large urban areas had established city-wide public/private sector regeneration partnerships which were evolving into powerful regional and city wide regeneration alliances. These networks centred around initiatives such as London Pride, the Newcastle Initiative, the Leeds Initiative, Glasgow Works, or Manchester Pride and included local authorities, key city businesses, and regional institutions such as Training and Enterprise Councils (TECs) and universities.

They formed a strategic consensus about a city wide 'vision' and a belief in strategic city and regional management and resource allocation. The vision may have been either speculative or vague with some partners more involved than others. But the output from these alliances was becoming quite impressive. Regeneration Strategy statements, flagship projects, major Lottery bids, and lavish expenditure on publicity and promotion were standard. Creating a business friendly image meant everything. In this context, although lip service to community involvement and meeting need was the justification for joint action, these needs had in practice a lower order priority than inward investment and corporate strategy.

Thus, corporatism (expressed through these strategic partnerships and other networks) was the conventional wisdom across the country well before the change of Government. UDCs were a form of corporatist urban development to the extent that business leaders, local authorities and government were represented on UDC boards. They were not, however, recognised as partnerships between public and private sectors, because they were government led and appointed bodies whose purpose was to specifically exclude certain local interests and to bypass local authority planning processes. Yet UDCs were not far away in structure and method from the partnerships that were to follow them.

City Challenge, and even more so Single Regeneration Budget (SRB), are explicitly developed around a three-way partnership model, with local authorities and representatives of the voluntary sector working with business and public sector agencies. Local authorities play a central role and there is less direct central government direction. Neither City Challenge nor SRB boards have powers to override local authority planning controls.

SRB has taken the partnership model further than City Challenge, requiring three-way partnerships to bid for funding and spreading the area of potential benefit across the country. Any area can bid for SRB, not only areas pre-selected by central government for urban funding on the basis of need. SRB is thus a more locally accountable partnership model, although 'local' in this sense generally means a local authority area or a large regeneration zone; it does not mean residential neighbourhood or local community. Moreover, the partnerships are generally made up of strategic 'players' selected or appointed by the key partners. They are not voted in or subject to public advertisement. Nevertheless, over the past five years, a substantial network of partnerships has developed which has proved successful in attracting government and European funding. In the fourth year of SRB, most areas have succeeded with at least one SRB bid.

Thus, given the existence of this structure of power, there is little pressure from the top to change the regeneration system. In fact, the opposite is the case; regeneration partnerships are looking to the Government to strengthen themselves.

From the perspective of the Government, reform of regeneration programmes is not high on the political agenda, compared for example with education or the NHS or the New Deal Welfare to Work. Significantly these other programmes interact strongly with regeneration, but regeneration itself is not a key national issue. It is a small budget in Government terms, it is relatively low profile politically, and is a 'local' rather than a national programme.

Nevertheless, there are voices urging change in regeneration policy. These include the Local Government Association (LGA) and the National Council for Voluntary Organisations Urban Forum and the Black Training and Enterprise Group and the Development Trusts Association (DTA). The LGA published prior to the Election a 'New Deal for Urban Regeneration' (now retitled 'The New Commitment to Regeneration') urging a stronger role for local government and for binding partnerships between central government and regeneration partnership boards, similar to the French *contrat de ville* arangements. The LGA is promoting a 'pathfinder' scheme to pilot their new model (LGA, 1998).

The Urban Forum and the DTA and Black organisations are urging a greater role for community development and public participation, and more effective targeting of programmes. The City 2020 Commission of Inquiry into Urban Policy (a Labour Party think tank) in its final report in April 1997, argued for an end to competitive bidding, with more democratic and open partnerships, capacity building funds for community led regeneration and much greater integration of regeneration funds with mainstream government funding (City 2020, 1997).

Some of the arguments for change are in potential conflict. Local government is seeking a greater role for local authorities, while some community organisations want less local (and central) government control of regeneration and more community control. While local authority associations are urging a strengthening of the strategic role of local authorities, the voluntary sector and ethnic minority organisations for example, want a strengthening, and better resourcing, of grass roots, neighbourhood level organisations (Thake and Staubach, 1993).

The appeal for reform from these different quarters has not fallen entirely on deaf ears. The first Minister for Planning and Urban Regeneration, Richard Caborn, issued new guidance for Round 4 of SRB very soon after taking office, stressing better targeting of resources on need and the importance of ensuring that SRB bids reflected the new Government's policies on getting people back into work and helping the disadvantaged. A later discussion paper, 'Regeneration Programmes: The Way Forward' (DETR, 1997), went further in encouraging more community involvement, better targeting, and an end to annual bidding rounds. An important emphasis was on integrating social and economic regeneration so that it was better coordinated and 'holistic'.

Thus, two levels of regeneration policy (and policy audiences) are evolving. At a strategic level, city wide and regional institutions are establishing themselves as key regeneration networkers in a top down way. This level is strongly reinforced by local government associations and also through the creation of regional assemblies and Regional Development Agencies (RDAs). At a local level, a more inclusive and needs based regeneration policy is being developed, this level being reinforced potentially by the linkage of SRB to local Welfare to Work programmes, special Action Zones (for health and employment for example) and by the Government's support for community involvement and 'holistic' solutions.

The two levels have evolved in different contexts; the strategic level reflecting the weakness of local government in the 1990s; the local level reflecting frustration of local communities with the powerlessness and lack of inclusiveness of local authorities. It could crudely be argued that the strategic level is primarily concerned with creating partnerships and ensuring coordination with the private sector about inward investment, while local level regeneration is expressing a more complex mixture of economic, environmental, and social objectives. The objectives of the two levels are not necessarily in conflict but there is a definite pecking order which requires the local level to be subordinate to the imperatives of strategic planning and the strategic economy.

In the community sector, there was also a complex picture which mirrors the two levels described above. First of all, some of the large voluntary sector organisations have been offered seats on both strategic and local partnership boards and accepted them in the belief that they can argue from the inside for a reallocation of resources and power. However, as several observers have pointed out the partnerships do not offer equal power or capacity to participate and for some voluntary groups disillusion has set in (see Mabbott, 1992; NCVO, 1994, Robinson, 1998).

Some of the smaller community based regeneration projects running training schemes, or volunteering programmes, have done reasonably well from SRB and City Challenge programmes, gaining just enough resources to run their projects. For others, expectations of gaining financial support from local authorities have been lowered due to year on year cuts to the voluntary sector and due to exclusion from the early rounds of SRB. Some turned instead to the National Lottery Boards or to European funding.

Other local groups have not become involved in the bureaucracy and 'beauty contest' of regeneration bids and partnerships. They have chosen to take direct action to change their environments (and to express their cultural preferences) through squatting, creating rave collectives, and forming anti-road and 'land for the people' occupation campaigns. In 1996, for example, an occupation took place, lasting until eviction three months later, of a prominent riverside site in Battersea which the owners, Guinness Ltd, wanted for luxury housing. The protesters, from the Land is Ours campaign, were demanding that the land be used to meet local housing need.

In Luton in 1992, the Exodus collective (a rave and New Age collective) occupied a derelict hospital site (owned by the local authority) and an abandoned farm (owned by the Department of Transport for road widening). Within a year the hospital site had been converted into housing for members of the collective and the farm into a community farm complete with animals and on-site staff to act as guides to visiting school children. This was achieved with no regeneration funding. Resources were raised from raves and dances and by pooling Housing Benefit. Indeed the local authority along with the police, and a spate of arson attacks, did everything to stop these 'direct regeneration' schemes going ahead. But after a series of court actions, the authorities have been forced to accept the reality of Exodus. The farm and the 'Housing Action Zone' are a model of grass roots initiative and imagination (Hart, 1998).

Elsewhere at the community level, very large numbers of people, particularly from the most disadvantaged communities and groups, are experiencing widespread exclusion or non-engagement in urban regeneration. As many of the ethnic minority organisations have observed, the most marginalised groups and individuals have gained very little from regeneration programmes and are rarely consulted or involved (Black Training and Enterprise Group, 1994). In fact, it can be argued that in spite of the rhetoric of equal opportunities within regeneration programmes, urban regeneration programmes have failed to deliver equal opportunities for ethnic minority communities on the ground (Loftman and Beazley, 1998).

As for the private sector, there is again a complex mixture of views. The large companies have come to terms with private/public partnership and are now enthusiastic, seeing the benefits, such as access to grant funding and influence over regional economic policy and local land use planing decisions. To take a crude example, with large supermarket chains playing their part in TEC training and New Deal Welfare to Work programmes, and by the creation of jobs at new superstores, these chains might expect to obtain planning consents for edge of town superstores more easily than if the superstores were judged strictly on their planning merits.

Thus, partnerships covering a range of economic and social concerns have enabled the private sector to exert much greater leverage over the land use planning system than in previous years. This goes well beyond negotiating planning gain agreements, to a major policy change where there is much greater flexibility in land use zoning and planning standards; an approach that mirrors what the UDCs were doing during the 1980s. Thus, public/private sector regeneration partnerships are achieving the same outcomes as the UDCs without their confrontational politics.

Smaller companies operating at a local level think differently from big business. Weighed down as they see it by business rates and high rents (in spite of being in poor areas of the city), with high level of crime and spiralling costs of insurance, many small firms can see little advantage in partnerships with local authorities and others. Too insignificant to be represented on strategic partnerships, and too hard working to have time to attend local meetings, small business generally feels left out.

Although some high streets have formed retailed associations and have, for example, campaigned for rates reductions or have bid for Home Office funds for Closed Circuit TV, they are small players in the regeneration game.

The conclusion to be drawn from this picture is that the 1997 General Election did not precipitate a significant change in the balance of power in regeneration, largely because the existing networks, particularly at the regional or city wide level, are working well for the key partners. Yes, the partnerships look for more resources and a closer working with central government, but they do not want an upsurge of people power or small traders' power to rock the corporate boat.

Economic imperatives

Whilst institutional factors are placing limits to change, economic imperatives are an even larger constraint on people-based regeneration and are reinforcing the layering of urban policy. Rising unemployment and industrial
restructuring, combined with reductions in government funding for mainstream services have established the priorities for economic development. The
imperative is to attract private investment, to restructure the labour market,
and to cut mainstream social programmes.

Thus, throughout the 1980s and 1990s, there was increasing competition
between towns and cities for private investment, and for public funding from
Europe or the UK. Reductions in UK government main programmes and
regional aid simply increased the pressure on cities and regions to compete
for private sector investment and for funds from Europe.

This investment is seen as crucial to create a new economic base for declining areas. In most cases, urban managers in the public and private sectors
have largely written off the indigenous economic base, particularly if it is
tied to manufacturing or transport, and are chasing footloose capital. Major
manufacturing companies are targeted in some regions, for example, car makers or micro-chip manufacturers, but the most common form of inward
investment is chain or specialist retail, leisure or business and financial services. Meeting the infrastructure, housing and training requirements of these
potential investors is the imperative for regional and city managers. Other
social, environmental and economic issues are of lesser importance.

Strategies for levering in investment are increasingly similar with local and
regional economic development campaigns following familar lines – promoting a positive image, advertising regional incentives, upgrading transport
infrastructure, reclaiming derelict land, compiling land and property regis-
ters, producing business competitiveness strategies, and marketing the result-
ing 'product' in the UK and overseas.

These factors do matter to business. At the margin, those towns and cities
that sell themselves well (and are able to offer the greatest financial incen-
tives) are able to claim increases in investment and jobs (for example the
Scottish Development Agency and the Welsh Development Agency). There
are formidable obstacles to 'local economic development' due to globalisation, economic cycles, and well established regional variations (London and
the South East versus the North East for example). Attracting inward investment can also bring new problems such as the downturn in Asian economies
now impacting on Asian investment in Wales and the North-East. Yet local
and regional marketing does make some difference and no local authority or
strategic partnership can afford to stay out of the competition.

This imperative has the effect of reinforcing the influence and the capacity of regional and city wide level agencies and partnerships. In fact, the drive
to create Regional Development Agencies is based upon the assumption that
regions must compete with each other both within and outside the country.

They define a regional economic vision, pool resources to fund marketing campaigns, and obtain ownership of how the region is sold (its image) and take control of the inward investment strategy.

Local authorities as gatekeepers

There is one key player who straddles both strategic and local partnerships, has considerable resources and powers of its own over the local economy, and is seen as a principal gatekeeper, by business and the local community. It is a gatekeeper that is divided between obligations to both the strategic and community levels, and between economic and social imperatives. That gatekeeper is the local authority.

Although the private sector is crucial to the credibility of partnerships in the eyes of central government, it is public sector bodies such as universities, health authorities, housing trusts, TECs and development agencies, which are the initiators and do most of the work of partnership building. They bring together public funding, land and property assets, and control legislative responsibilities such as planning, housing and environmental health which are critical services for regeneration. But they do more than that. Local authorities are the key brokers in many strategic regeneration projects because they can talk to all sides, they have the legitimacy to do so, and have crucial influence over grants and the legal responsibility for making development control decisions (see Imrie and Race, 1999).

Critically, there is significant overlap of representation on boards, committees and councils. This overlap has been developing for some time and is very well established in some regions. Shaw *et al.* (1996) have written about Quango membership in the North-East of England. Senior councillors pay a key role in these organisations. This public sector network is no longer an Old Labour network. It is leavened by senior private sector and public sector agency appointments and is becoming much more entrepreneurial.

The private sector must be 'on board' because large businesses, with significant local investments, create jobs, train local people, and act as magnets to attract other investment. Moreover, the private sector brings in 'match' funding against which grants can be levered in from Europe, and also in kind contributions and project management skills. In other words, the private sector provides the key 'outputs' from regeneration that the government is looking for. There is also an increasing interdependence between sectors in delivering a range of quasi-social programmes such as Care in the Community, Training, Welfare to Work, social housing and Health Action Zones. The outputs here are not conventional private sector leverage but the private sector is involved because they are bidding for service contracts where the client is the public sector.

There are great dangers in the ever closer links and joint working between local authorities, other public bodies and private investors. Boundaries between public and private get blurred. Public representatives, including elected local councillors, can lose sight of their public accountability and their duty to the public realm. The private sector may assume that the public sector can in some ways be 'bought' by entertaining and planning gain deals. Press stories have come out of cross overs of public and private business arrangements (for example the Welsh Development Agency and Doncaster Council), which suggests that public/private sector partnerships are in need of scrutiny by the Nolan Committee – and more attention by the press.

These arrangements are often hidden from the public eye. They operate in secrecy without published minutes or agendas with meetings that are rarely open to the public. This secrecy is in the name of confidentiality about negotiations for contracts and land deals for major projects. But there is a democratic price to be paid. As Loftman (1998) points out, 'The secrecy which often shrouds negotiations concerning the projects further serves to constrain public debate'.

This issue of transparency is especially important because the local authority has a critical leadership role, as broker and champion – and this role is increasing in importance. It is, thus, faced with an increasing contradiction between democratic legitimacy (see below) and secret deal maker.

John Stewart has written that for local government to have a future it must provide 'community leadership' – not just leadership in a narrow party political or neighbourhood sense – but in terms of bringing together a wide range of interests across a local area or region. It must create and lead a common agenda (Stewart, 1997). The government has now adopted this agenda with enthusiasm. It forms part of its strategy for modernising local government (DETR, (1998a, 1998b).

Significantly one of the reasons for promoting the community leadership role is precisely because there is a crisis of legitimacy of local government. Turn-outs at local elections are low, there is little belief that local councillors are sufficiently representative, some councils (though not all) deliver poor services and generally there is a 'them and us' relationship between local residents and businesses and the Town Hall. In this deteriorating climate of legitimacy, reinforced by the reductions of powers and resources to local government throughout the 1980s and 1990s, local authorities have to work hard and creatively to regain legitimacy as community leaders.

Central government is known to be highly sceptical of the ability of local authorities to take this leadership role. A study by INLOGOV showed that residents trust their own local associations more than central or local government, and that councillors and local authorities have a lot of work to do to regain credibility. It recommended that councillors should develop political networks outside traditional party structures to cultivate community leadership (Hall, 1997).

Community leadership operates at both a strategic level through city wide partnerships but equally at the level of local area partnerships. Local area regeneration is unlikely to get far if it is not supported by the local authority. Most local authorities produce regeneration strategies or statements for local areas that set out priorities and key issues. These area strategies will determine funding priorities, and bids and partnership structures. Most SRB bids are led by local authorities or are strongly influenced by them.

Thus, projects are unlikely to get off the ground without local authority support. Officers from the local council are often crucial to facilitating local project development. The Town Hall will always have many more full time officers working on regeneration than either the voluntary or private sectors. Moreover, the Town Hall will have its hand on many of the key regeneration levers – town planning control, land ownership, access to grants. The Town Hall will also be fully networked into the Government Office for the Regions who determine bids, dispense SRB monies and monitor performance. Thus access to the local authority at many different levels and across many different departments will be crucial for the success of all projects whether they be intiatited by the public, private and voluntary sectors.

It follows that the exact party political balance, ideological stance, and the political and officer culture (consultative, open or defensive for example) and the resources put in by the local authority to partnership building or community development will be critical factors in local regeneration. It may be possible for some projects to be successful without this support, but it is very unlikely unless there are powerful allies elsewhere who can override or bypass local politics.

The gatekeeper role can thus be summed up as follows:

- local authorities set the local policy framework for regeneration but always in a way that is flexible enough to respond to the 'big offer' if it comes along;
- most projects need local authority support to get central government or Lottery funding;
- local authorities largely control the partnership structure which determines strategic priorities and the allocation of SRB and ERDF funds;
- local authorities act as the key broker on strategic deal making;
- local authorities own key sites that are targeted for flagship developments;
- local authorities control the local planning system which determines whether new developments are allowed and although this should be independent of their views on regeneration funding, it does have an influence on resource allocation.

The gatekeeping role is, however, now facing new tests. Precisely because of the lack of access to SRB which many community groups are finding, such as the difficulty of getting included in local authority run SRB programmes, some community organisations are taking things into their own hands and putting in their own SRB bids, thus, effectively bypassing the local authority and appealing directly to the Government Offices for the Regions. The question is whether Government Offices will ask local authorities for their views on these bids, in other words, find out whether the local authority supports the bid, or whether Government will consider these bids on their merits.

In, in summary there has been political change in central government with a flurry of new initiatives, but perhaps little change in the control of regeneration and the outcomes for the following reasons:

a. In spite of the new language of inclusion and partnership, economic competition remains the primary driving forces in regeneration so that the economic and social outcomes of regeneration have not significantly changed.

b. The strength of strategic/regional partnerships and their influence over strategic deal making mean that there has been continuity in the network of urban and regional managers in control of the process of urban development.

c. Grass roots involvement in regeneration has been effectively reduced to a minimum by the gate keeper role of local authorities who control access to SRB programmes, allocation of regeneration funding, disposal of land and assets and town planning control.

Thus, local authorities are faced with, on the one hand, pressures from the grass roots to allow more people into the regeneration process through greater consultation and local participation. On the other hand, they are acting secretly and less inclusively in their strategic brokering and deal making role.

This conflict was missing during the era of UDCs since UDCs were given the 'single minded' task of regeneration in Urban Development Areas. They were able to ignore grass roots views and were encouraged by Government to act behind closed doors. These arrangements and assumptions no longer exist. Local authorities, whether they are inside or outside partnerships, are not like UDCs. Central government itself is asking local authorities to be more open and democractic in its dealings with the public and business. Thus, a new politics of regeneration has emerged which local authorities are now wrestling with.

The question to be addressed below is whether this new politics also creates opportunities for grass roots regeneration that did not exist before. Can the regeneration process be opened up so that the balance of power is changed?

The opportunities

It is argued here that in spite of the elaborate strategic level networking which effectively excludes grass roots and previously marginalised groups and influences, there are opportunities for people based regeneration particularly at a local level in a way that was not possible in recent years.

At a strategic level, it is not an exaggeration to say that regeneration policy is largely 'sewn up' for the reasons discussed above. Local groups are unlikely to get access to the strategic level of policy making. But the language of inclusion of the new government (and the European Union) added to the policy vacuum about regeneration at the local (neighbourhood) level, means there is now developing a major battle for control of locality agendas. Neither strategic partnerships nor local government in its 'community leadership' role (though there are some exceptions) have yet determined how local area regeneration and community development could or should take place nor how this fits into wider strategic priorities.

Many local authorities are wary of area based devolution initiatives. They are also worried for political reasons about recreating Community Development departments and funding community development officers. Some have dived in and created Local Area Partnerships backed up by community development strategies (for example, Liverpool City Council, and Sandwell MBC) but these are exceptions. Thus, the form of neighbourhood regeneration, the process of participation and policy decisions about the allocation of resources are all to some extent up for grabs.

This locality opportunity appears to contradict the argument of the earlier sections that regeneration was 'sewn up' by the partnerships and local authorities. Given that the same factors of partnership power relations, economic imperatives and gatekeeping by local government apply at the local level, how can localities be an independent force?

First, private sector partners and major public agencies do not need to be concerned with neighbourhood politics partly because they expect that to be taken care of by local authorities. Secondly, community groups have generally beeen unable to prevent strategic developments they do not like. Thirdly, an increasing strategic consensus around economic development leads to some complacency by strategic partners about neighbourhood level politics by all strategic partners including local authorities.

Fourthly, while economic factors are critically important at all levels, there are likely to be more 'not for profit' initiatives (e.g. social housing, community economic development, voluntary sector activity, volunteering) in neighbourhood regeneration which have a relatively independent life. The local authority sector as we have indicated has very divided loyalties between strategic partners and local needs and its electorate. In this vacuum, there are openings for grass roots development whose success will depend on the precise nature of local politics.

In summary, therefore, we can indentify the following opening for people-based regeneration;

(a) The increasing emphasis on 'need' in Government regeneration guidance, places more pressure on partners in regeneration to deliver on real benefits. In other words, instead of 'bypassing communities' as UDCs were accused of doing, regeneration partners are under pressure to include these communities in their programmes and to target resources at them. Partnerships can perhaps be held accountable to this objective.

(b) Expanding the community leadership role of local authorities need not solely advance the power of local authorities to pursue their own corporate agenda but could lead them to be more innovate, creative and inclusive. In the search for greater legitimacy, local authorities can take more 'risks with democracy' if they wish to and thus increase community empowerment. Experimental arrangements for democractic accountability including elected mayors, referendums, and public scrutinies may have this effect. It is possible, of course, that some of these innovations, in particular elected mayors, may have the opposite effect of weakening local democracy, but the fact of a debate about local democracy challenges the legitimacy of existing institutions and opens up the possibility of new ones.

(c) With local authorities somewhat distracted with bidding for funding and investment through strategic partnerships, neighbourhood based regeneration and action by marginalised groups and communities may begin to have some effect.

(d) If local groups can apply for funding directly to Government at regional or national level (which seems probable under new regeneration guidance) they may be able to bypass the local authority if it is acting as a restrictive gatekeeper. This could have the effect of freeing up resources for the grass roots, and liberating grass roots ideas that may not fit into the local authority way of thinking.

(e) If local authorities themselves become more entrepreneurial, thinking creatively about their roles and the services they provide, as many are doing, they may be required to pay closer attention to the local economy, local people and their needs and skills, and cultures, thus reducing their dependence on higher level partnerships and central government.

(f) With the shift to local governance (a recognition that the local authority cannot do everything), there is an increasing awareness of the need to spend public and private resources on 'capacity building' in both the small business and community sectors so that these sectors are able to come forward with projects and ideas, and be active partners in policy development and service provision.

The New Deal for young people and the unemployed may have some impact at an individual level, but unless it is complemented by local area and neighbourhood empowerment, it will not get very far in transforming the economic and social fortunes of deprived areas.

There is a vacuum in neighbourhood regeneration that is there to be filled. The present uncertainty about the role of local authorities and the contradictions in the partnership approach mean there is no clear strategy for neighbourhoods or for marginalised groups. Local groups have everything to play for.

Further reading

A useful read on community regeneration is City 2020 (1997) while readers will find Thake and Staubach (1993) a good reference to refer to. Oatley (1998) provides a broader assessment of recent changes in urban policy towards a more user-sensitive approach. It is also worth perusing the journal *Local Economy* which often features articles on community regeneration.

11

Just another failed urban experiment? The legacy of the Urban Development Corporations

ALLAN COCHRANE

Introduction

The short history of British urban policy is strewn with the wreckage of a series of experiments and initiatives. Just a brief list should be more than enough to give a flavour of this: from Community Development Projects, Inner Area Studies and Comprehensive Community Programmes to Enterprise Zones, Task Forces, Urban Development Corporations and City Challenge; from Urban Programme and Inner City Partnerships to the Single Regeneration Budget and URBAN. And that is without taking into account initiatives in the fields of housing and education. Each new policy fad and ideological twist of government policy seems to be given an urban dimension. Each new initiative is presented as an experiment, or a new departure, on the basis of which lessons will be learned and policies generalised. In practice, however, the main lesson of Britain's urban experiments seems to have been that by the time each one has come to the end of its life it's already time to move onto the next, stepping over the failures of the past into a bright new future (see, e.g., Atkinson and Moon, 1994; Burton, 1997; Edwards, 1997; Edwards and Batley, 1978; Higgins *et al.*, 1983; Loney, 1983 and Wilks-Heeg, 1996).

It would be difficult to trace a very clear lineage between all of these different initiatives. As Edwards (1997, p. 832) notes, 'Every urban policy we have put in place since 1967 has been transient. Each and every one has been a special programme, either with a fixed life of its own or providing funding for a fixed period.' But this has merely encouraged policy analysts in the search for an underlying logic. The dominant orthodoxy suggests that there was a sea change (or 'watershed', Atkinson and Moon, 1994, Ch. 4) associated with the 1977 White Paper on the inner cities (Department of the Environment, 1977) (see also Lawless, 1989). The White Paper is generally presented both as the most coherent official analysis of the inner city problem and as confirming a shift away from an approach that focused on the social pathology of urban 'communities' to one that drew on structural explanations. Instead of blaming the residents of the inner cities for their predicament, the explanation was to be found in broader economic shifts, and the impact of economic restructuring. The policy emphasis shifted from social to economic regeneration. The post 1977 period, before the advent of full-scale Thatcherism, has sometimes been interpreted as a brief 'golden' age for urban policy in which social and economic factors were explicitly linked together in the development of strategy (e.g. Wilks-Heeg, 1996).

But this may exaggerate the significance of the shift. It certainly exaggerates the clarity of shared vision, seeking to construct a unitary 'urban policy' where none existed in practice. By contrast Edwards (1995) suggests that the 1977 White Paper and the legislation that followed actually made it more difficult to develop an urban social policy. Higgins *et al.* (1983) are also scathing about the White Paper's use of the language of co-ordination and partnership which they argue was a substitute for the development of a rounded analysis of urban problems. In practice succeeding urban policies have been concerned with different issues, even if they share the 'urban' label – from 'race' to community breakdown, from small firms and a lack of entrepreneurial spirit to economic decline, from environmental degradation to crime and drugs, from unemployment to lone parents, from inner city to peripheral council estates. A scrutiny of the published aims of urban policies by Robson *et al.* (1994, p. 5) confirms that they have had 'well over 100 programme objectives'. It is often unclear why some policies attract the 'urban' label, while others do not, despite their significance for people living in cities (Blackman, 1995).

So, is there anything that holds urban policy together?

Unlike other forms of social policy, of course, urban policies are area based. That is, they focus on areas whose residents, economies or environments are defined as facing problems of deprivation, decline or degradation. They are not concerned with the delivery of services to people on the basis of some more or less agreed (universal or targeted) set of rights, entitlements or conditions. It is not even generally the case that all those living within the designated areas are entitled to particular forms of welfare benefit or access to particular services, although in the past the designation of general improvement areas or housing action areas has made this possible for narrowly specified purposes. Urban policy expenditure instead tends to flow through particular projects and programmes, following a process of selection based on a changing set of criteria, which are rarely specified with any precision.

More tentatively, as Robson *et al.*, (1994) suggest, it is also possible to argue that this area focus means that urban policies may be able to take a more 'holistic' approach to social problems and ways of tackling them. Instead of complaining that such a wide range of apparently disparate issues have attracted the 'urban' label, maybe it is more appropriate to celebrate the possibility of linking them together. Most of the initiatives seem to have incorporated a belief in the value of co-ordination between agencies, or the notion of an integrated approach. Although the significance of this should not be exaggerated there is also some sort of shared understanding that there are linkages between social and economic factors, even if the implication of these linkages is rarely explored explicitly. The policies which start by analysing urban problems in terms of the social pathology of residents promise to improve local economies by strengthening communities (encouraging moves away from welfare towards work) (see, e.g., Thake and Staubach, 1993), while in their more structural phases urban policies seem to incorporate a belief that economic regeneration will also lead to improved social welfare for local residents, making 'inner cities places where people wish to live and work' (DoE Circular, quoted in Higgins *et al.*, 1983, p. 83; see also Robson *et al.*, 1994).

Another – rather unfortunate – shared feature of urban policies in Britain is that the widespread academic and professional consensus seems to be that, in terms of their stated or implicit aims, they have all been rather unsuccessful. They have neither significantly improved the social and economic position of those living in the inner cities and peripheral estates, nor have they succeeded in delivering the hoped for economic renewal. Some commentators are more harshly critical than others, but they do, at least, all seem to agree that little progress has been made. Some blame the short-sightedness of policy-makers, others the intractability of the problems. Some blame the incoherence of the aims, others the lack of co-ordination. Some simply think the levels of expenditure are inadequate for the tasks in hand (see, e.g., among many others, Atkinson and Moon, 1994; Audit Commission, 1989; Burton, 1997; Edwards, 1995 and 1997; Edwards and Batley, 1978; Lawless 1996, Robson *et al.*, 1994).

How do the Urban Development Corporations fit into this history?

In some respects, at least for a policy analyst, the experience of the Urban Development Corporations is reassuringly familiar. Like many of the other initiatives, they were launched on a sea of optimism. The promise was that the Urban Development Corporations and perhaps the London Docklands Development Corporation, in particular, would generate dramatic change in the inner cities, transforming areas of dereliction into beacons of hope, through an integrated approach to development. Like earlier products of urban policy, they were area based projects located in a limited number of places. The 'problem' was defined in terms of areas, rather than people. And there was never any suggestion that every urban area would also have its own Urban Development Corporation. As with other initiatives only some areas were chosen as suitable for treatment. The places chosen were not necessarily the ones in the worst condition, nor necessarily those likely to be most susceptible to intervention, where intervention was likely to be most successful. Like other initiatives the criteria on which particular areas were selected were never made explicit.

On balance, too, just like the other initiatives, most commentators seem agreed that the Urban Development Corporations failed as a 'solution' to urban problems (see, e.g. Atkinson and Moon, 1994, pp. 143–154; Imrie and Thomas, 1993b; National Audit Office, 1990; Robinson, 1989; Turok, 1992). Despite some sympathetic comments, Robson *et al.* (1994, p. 54) confirm that there has been little 'trickle down' from initiatives such as the Urban Development Corporations to local residents. Even if the emphasis is placed on the attempts of the Urban Development Corporations to generate or encourage private sector development, rebuilding confidence among investors, their success has been limited and achieved at financial costs far greater than initially predicted. London's Docklands has probably been most dramatically transformed, but not only has infrastructural spending by the LDDC probably been greater than that of all the other development corporations put together, but it leaves a legacy which will require continued massive capital expenditure (e.g. in the building of roads and the extension of the Jubilee line) (Edwards, 1995, p. 699).

Today it seems that the Urban Development Corporations, like the enterprise zones before them, are being allowed to sink relatively quietly into the sunset, while the rest of us are left to wonder what all the fuss was about. No new ones have been set up since 1992, and there is no suggestion either from government or the policy community that the time has come for the launch of a whole new set of Urban Development Corporations. So, it would be easy simply to dismiss them as just the latest in a continuing series of failed urban experiments.

But such a conclusion would be premature. The Urban Development Corporations also incorporated some quite distinctive ideological twists, which deserve further consideration. Although land assembly and the reclamation of derelict land have been integral to inner city policy for a very long time, the corporations defined the urban – or inner city – problem almost entirely in terms of land and property, dereliction and a lack of development, rather than poverty, unemployment or even, strictly, economic decline. Their policy emphasis was on renewal through private sector led property development. This represented a break, even with previous regeneration focused initiatives which stressed the renewal of manufacturing and related industries. The focus on property development meant that success could be measured in terms of the construction of buildings and their occupation at market rents. If anything, attempts at the attraction or maintenance of traditional industry might interfere with this strategy, making it more difficult to market property to the new growth sectors in services and high tech industry.

Byrne (1997) notes the extent to which Urban Development Corporations and related policies can be seen as part of a wider attempt to reinforce the depoliticisation of planning, encouraging marketisation as an alternative and seeking to create a 'real' market in inner urban areas. The extent to which the Urban Development Corporations succeeded in encouraging the restructuring of urban property markets may be questionable, but they have acted as 'flagship' projects in the redoubts of 'labourism', contrasting the vibrancy of service-led markets with the stasis of declining manufacturing and industrial areas. They helped to confirm a move away from explicit concerns with social welfare, instead offering the promise of trickle down from economic success. Development and the revalorisation of previously apparently worthless land were themselves seen as measures of success. This powerful ideological message helps to explain the readiness of a 'free market' government to keep investing more.

The flagship Urban Development Corporation in London's Docklands acted both as a fundamental reproach to the Labour authorities of the area, and – more important – as a powerful image to the rest of the UK of the success of Thatcherism in its heartland (see also Allen *et al.*, 1998). Although the imagery was less dramatic, the other development corporations offered similar messages. In that sense they were the urban element of the wider programme of political and economic restructuring associated with Thatcherism (see also Anderson, 1991).

Unlike the Enterprise Zones, however, the Urban Development Corporations were not pure children of Thatcherism, because they drew on a clear cut interventionist agenda. It was always clear that they would require large sums of state expenditure to achieve their aims – they were explicitly interventionist. They were, in other words, the institutional children of Heseltine as much as Thatcher. They were initially associated with a charismatic politician for whom they represented a new political model of business-led purposive state intervention. Despite the rhetoric of planning deregulation, they represented a massive state investment, and, where it was deemed necessary, they were prepared to use some of the most interventionist forms of town planning – including compulsory purchase orders – to achieve their ends. Raco (1997) highlights the lack of sympathy given to small businesses in the face of the drive to restructuring and modernisation. In Chapter 2 of this book, Brownill notes the negative implications of defining Docklands as a 'greenfield site' for the wide range of already existing local businesses (see also Brownill, 1990, pp. 96–9). The Thatcherite agenda made it impossible directly to espouse an interventionist industrial strategy, but the Urban Development Corporations offered a different route to interventionism.

The Urban Development Corporations have been criticised for their part in encouraging a move away from elected local government in the context of a wider range of policies directed at reshaping forms of local governance in the late 1980s and early 1990s (see, e.g., Cochrane, 1993). There are three main ways in which the corporations incorporated this agenda. First, local structures of democratic accountability and bureaucratic organisation were explicitly identified as problems by Heseltine and others (see Chapter 1) because they made it more difficult to follow through an effective, coherent and integrated programme of regeneration. A more business-like approach was needed. Second, business involvement on the boards reflected an ideological desire to undermine and question the legitimacy of elected local councils. Instead of elected councillors and their officers, the aim was to create a public status for representatives of business, as part of the process of creating a more entrepreneurial state. Third, since the boards of the Urban Development Corporations were appointed by central government, there was also the prospect of increased centralisation, with the corporations acting as agents of central government and little scope for developing programmes which reflected the needs of local residents, or even local businesses.

One of the distinctive purposes of setting up the Urban Development Corporations was to show the value of single-purpose agencies with a proactive entrepreneurial agenda. Such agencies would, it was argued, be able to overcome the inherent weaknesses associated with the divisions created by the overlapping jurisdiction of local government and other public organisations. In other words, they were intended to be examples of post-bureaucratic (focused, task-oriented and businesslike) organisations with a commitment to market-led solutions. They were expected to solve the perennial problem of co-ordination between a range of different agencies, each with their own 'interests', by integrating them into one organisation. The corporations were given the sole responsibility for a range of development activities – to cut the bureaucratic knot associated with local government attempts to co-ordinate and work together. Like the new town development corporations they had dedicated professional staff who led and developed the agenda, under the nominal leadership of a business-led board. According to Robson *et al.* (1994, p. 52) one of the strengths of the Urban Development Corporations 'has been [their] scope to develop more integrated programmes involving training, job creation, environmental and infrastructural improvements'. Paradoxically, for initiatives presented as businesslike, they did not have the same requirement as the new town corporations to generate income from development to cover their costs (nor, of course, did they incorporate the expectation central to Ebenezer Howard's garden city model that any gains would be fed back to local communities).

Many of the early fears raised by the critics of the Urban Development Corporations were exaggerated, even if they cannot be completely discounted. Despite a shared emphasis on property development, the policies of the Urban Development Corporations were neither monolithic, nor fully determined by central government. Each developed a rather different approach to its work. As the other chapters of this book confirm, there were substantial differences between the corporations, and their relationships with other local agencies varied significantly. The size of the different designated areas ranged from 187 to 4,858 hectares, while the nature of the areas differed nearly as starkly, even within the same conurbation in the cases of the

Central Manchester and Trafford Park development corporations. In some cases there were significant existing local populations, in others there were very few residents. The levels of funding varied dramatically, too.

In practice the Department of the Environment found it increasingly difficult to control the expenditure of the various development corporations (particularly, but not only, the LDDC), and this may in the end have helped hasten the end of the 'experiment'. Wilks-Heeg (1996) explores the way in which the LDDC was able to ride the rising London property market of the mid-1980s. The generation of income from property sales allowed it to extend its ambitions towards major redevelopments like Canary Wharf. But another consequence was that the government was forced to provide much more public finance to underpin LDDC's more grandiose schemes, particularly in the wake of the property slump at the end of the 1980s and into the 1990s. The vagaries of the London property market have created difficulties for Docklands, but at least there does seem to be a continued basis for expansion based on the relocation of City related functions. Outside London, matters are still less clear cut, as Urban Development Corporations had to reshape their strategies in the face of declining property markets (National Audit Office, 1993). Despite the early rhetoric, the corporations turned out to be more effective in developing linkages with public sector agencies, and better at levering grants from government for the development of infrastructure, than at 'levering' large scale investments out of the private sector. They not only called on a dramatically increasing share of identifiable inner city expenditure in the early 1990s (around 57% in 1992) but also undertook a major part of the spending identified under other headings (such as derelict land grant) (see also Wilks-Heeg, 1996, pp. 1267–69).

To a greater or lesser extent urban development corporations had to develop a local base, to develop locally based alliances, and – even – in some cases to sponsor 'community based' organisations with which to interact, consult and negotiate. In most cases – despite the initial rhetoric (and with the notable exception of the London Docklands Development Corporation in its early years) – they even worked, formally or informally, with local authorities and their officers (see, e.g., Imrie and Thomas, 1995). Perhaps this is not surprising since staff were often drawn from the same pool as those working in local authority planning and estates departments, and in some cases there was significant movement between the locally based organisations. Unlike some of the new town development corporations, which had worked with largely greenfield sites and weak local organisations and councils, the Urban Development Corporations were inserted into a space rich in organisational networks, and often with strong local government institutions.

So, what is their legacy?

Just like the new town corporations (created by social democracy) the Urban Development Corporations (created by the new right) are now being relegated to the history books. Before we finally wave them good-bye, however, it may be worth thinking more seriously about the role that the Urban Development Corporations have played both in the development of urban policy and in shaping forms of urban governance. Have they helped to clear the way for different understandings, or new approaches?

Despite the waves of criticism directed against them (and the single purpose model), particularly from within local government, the Urban Development Corporations also represented a break from previous forms of urban policy because of the extent to which they took the (admittedly redefined) 'problem' seriously. Instead of believing that working together (or improved co-ordination between agencies and more effective partnership between the public, private and non-statutory sectors) would solve everything, the assumption was that a dedicated approach underpinned by powers of intervention and relatively large budgets was required. Government support was only withdrawn when the full costs of the approach became clear in the context of a falling property market. In other words, the rejection of the single purpose model may also be a rejection of policies which require governments to allocate significant resources to identifiable urban initiatives. Local authorities have been able to reclaim a leading role in the pursuit of urban policy on the more or less explicit assumption that they will not call on any additional mainstream (or even very much ear-marked inner city) funding.

It is important to recognise that, although the urban development corporations were not themselves partnership agencies, one powerful legacy has been to encourage the growth of public-private partnerships, as each local authority has done its best to show that it can achieve what an Urban Development Corporation can in partnership with local based businesses and other agencies. In the case of Sheffield, for example, a pre-emptive strike was taken by the City Council to avoid the 'need' for a development corporation (through the setting up of the a public-private-community partnership economic regeneration committee and the development of plans for development in the Lower Don Valley). Although a development corporation was designated in 1988, the language of partnership survived. Similarly in Birmingham, although a development corporation was designated in 1992, Heartlands was first set up as a local, rather than a central government initiative (Lewis, 1992, pp. 53–4). The argument of local government was that local initiatives meant that those involved had a better understanding of what was required, and that it was also possible to integrate a wider range of activities (including social housing) as well as reflecting community needs more widely.

Paradoxically, by rejecting an explicit partnership model involving local authorities, in the style of the old Inner Urban Areas Act, the creation of the urban development corporations challenged councils to respond. The ideological basis of the development corporations implied a redefinition of welfare in terms which stressed business and economic prosperity – and the architectural symbolism of property development – as the real measure of success. They showed in practice that it was possible to have a social policy which owed little to the direct provision of welfare services. They highlighted the possibility of an approach based on partnership with private sector agencies. They were symbols of managerialism rather than democratic accountability. In all these respects, of course, the development corporations fitted well with wider shifts in the workings of urban governance. But they also helped shape those changes. Their very existence and the implicit threat that the model might be generalised encouraged councils to adopt similar approaches and reinforced moves towards more entrepreneurial local welfare regimes. The City Challenge initiative, which succeeded urban development corporations in the history of urban policy, can be seen as a form of 'probation', forcing councils to offer signs of good faith in order to get central government funding.

The legacy of the models of governance associated with the Urban Development Corporations has been highly ambivalent. The spread of local(ist) institutions has also involved a growth in the numbers of people being appointed to boards of one sort or another by various Secretaries of State (sometimes endorsing the self-selection recommended by the boards themselves) (see, e.g., Davis and Stewart, 1993; Greer and Hoggett, 1996). This has operated at the heart of the welfare state (e.g. in the case of NHS trusts and Health Authorities) as well as in more distant agencies such as Training and Enterprise Councils. But the nature of the appointments has been less narrowly restricted than in the case of the urban development corporations (with a greater desire to involve wider community representation and a more explicit role for local government as nominator/participant). Again, therefore, they can be seen to be part of a wider shift without the particular activist (and interventionist) model of the urban development corporations being easily generalisable.

Although the organisational form of the corporations was inherently temporary it left a legacy as a demonstration project. This was reinforced by the explicit requirement placed on the corporations to produce 'exit' strategies as they prepared to wind down their activities. They were specifically asked to consider the issue of local 'capacity building'. By the time of their demise the development corporations were expected to have generated a range of wider local organisational and institutional legacies, as well as legacies expressed in the language of concrete and steel. It would be possible to take a cynical view of this move by suggesting that it merely confirmed the failure of urban entrepreneuralism as the basis of a renewal strategy. But it may be more significant than this. It implies a shift in policy emphasis towards looser forms of governance through a series of state, community, not-for-profit and private sector agencies. It seems to promise social orchestration through networks and partnerships, rather than state initiative, even if the precise mechanisms remain unclear. An emphasis on the development of local institutions and the linkages between them (maybe even 'institutional thickness') (see, e.g., Amin and Thrift, 1995) seems to be the next 'big idea', offering the prospect of integrating new forms of urban governance and urban policy. Similar requirements are now included in the briefs of all the main urban programmes (including City Challenge and the Single Regeneration Budget), although Edwards (1997, p. 833) comments that the practice is likely to 'fall short of expectations – particularly as the latter have been couched in the new 'managerialese' that serves to obfuscate rather than to clarify'.

Some straightforward policy legacies have also survived. The flood of image building and the commitment to prestige projects as symbols of renewal has not abated. The urban development corporations have been associated with what Edwards (1997, p. 826) has described as a 'new urban glamour policy' – the LDDC had Canary Wharf, Cardiff has plans for an Opera House, Liverpool had Albert Dock and the Tate Gallery. This association with prestige projects is an ever present element of contemporary local regeneration strategies, and clearly has little directly to do with the 'traditional' notions of inner city policy – see, e.g., Loftman and Nevin (1996). And it remains a significant element in national urban policy, for example in the competition over the siting of the Millennium Dome, which (like the first wave of Urban Development Corporations) is directly associated with a senior government minister. The use of National Lottery money has helped to reinforce this trend by providing a source of funds for major capital investments, without offering continued revenue support. Place marketing has been institutionalised with the launch of initiatives such as City Pride. There has been a shift in emphasis away from dereliction, decline and decay, towards one which stresses the cosmopolitan potential of urban areas.

The emphasis on a more or less competitive bidding process was extended into the City Challenge programme, and has now been further institutionalised in the operation of the Single Regeneration Budget. The introduction of the Single Regeneration Budget brought the incorporation of a substantial number of other pre-existing programmes (around 20) including the Urban Programme. The SRB is both project based (along traditional lines) and still more competitive even than the old Urban Programme. Competition between bids, rather than any overall programme of or strategy for intervention, becomes the way of deciding where state funding should be allocated. Edwards (1997) powerfully criticises this approach because it is incapable of taking account of social need, concluding that 'it is ultimately demeaning for those whose well-being depends on the results' (Edwards, 1997, p. 841). Despite these criticisms, the method seems to have become increasingly embedded in the operation of British social policy, as, for example, education authorities are invited to bid for extra resources for areas whose schools face particular difficulties.

The urban development corporations were one element (among many) in the wider restructuring of the British state which now makes it difficult to go back to the old innocence of the Keynesian/Beveridgean welfare regime.

The imagery of urban renewal represented by the prestige developments of London's Docklands dominated the late 1980s and early 1990s as a symbol of what was possible – from wind-surfing on the Thames to the relocation of Fleet Street. Although some of the development corporations outside London had less to show, they too implied that it was not just possible, but also necessary, to buy into this new set of visions (from Cardiff's Opera House, to a whole set of waterfront developments). It is now a part of the dominant political commonsense that welfare stems from economic and entrepreneurial success, rather than a system of social security and welfare benefits. The experience and promotional activities of the Urban Development Corporations have played a key part in encouraging this shift.

The final twist

The emphasis of urban policy over the last twenty years has tended to be on economic regeneration. The dominant assumption seems to have been that if the local economy could be sorted out, then local communities would benefit, too. The approach pursued by the Urban Development Corporations was consistent with this emphasis, because of their stress on the importance of development. But they went further than previous initiatives because they started out with little interest in local populations – in several cases, regeneration was also expected to change those populations by bringing in new ones.

The justification for setting up Urban Development Corporations, however, rejected the old structural explanations of inner city decay. On the contrary the implication was that the state of the inner cities was a more or less direct consequence of Britain's welfare culture and the politics associated with it. The problem was defined as one which arose from discredited social and political arrangements. Planners were insufficiently entrepreneurial. Councils were more concerned to defend their own interests instead of being prepared to work with others, particularly private developers but also other government agencies. They stopped the market from working efficiently – clearing the inner cities and rebuilding them in another form (see, e.g. the discussion of market-led planning in Brindley *et al.*, 1989). They were more concerned with delivering services to passive recipients than with providing their 'clients' with routes out of deprivation. It was in their interests to maintain people in dependency, because it justified the existence of their departments and helped to generate their budgets. In a sense, this was an argument which blamed the problem on the social pathology of the local welfare state. Self activity and private sector investment were presented as the alternative to dependency.

The demise of the urban development corporations has been accompanied by a rise in policies which have begun to redefine the problem once more in terms of the people who live in the inner cities. There is now a renewed emphasis in social policy on problems of pathology on notions of dependency and ways of challenging it. The answer is presented in terms of a move from welfare to work. Training, rather than local economic development, has become the panacea of the 1990s, alongside the (re)discovery of 'community based' initiatives (see, e.g., Thake and Staubach, 1993). Local authorities all over the country have begun to focus attention on area based anti-poverty strategies. A social exclusion unit has been set up by central government. The language of national urban policy has swung to partnership, co-ordination and working together, and away from the economic regeneration of the inner cities. The point in the urban policy cycle seems to have shifted back to pathology, back to poverty, and back to social policy – even crime has become a target of intervention once again, with localised proposals for curfews and policies of zero tolerance for all criminal activity.

Wilks-Heeg (1996) sees the urban development corporations as a dead end rather than the basis of a new approach to urban policy. He highlights (and supports) a move back to an approach which has a more balanced social/economic agenda, reflected in the priorities of the Single Regeneration Budget and the 'partnership' approach implied by City Challenge and City Pride. In other words, for him, the lesson of the urban development corporations is that the social – or community – element cannot be removed from urban policy if it is to improve the position of those living in deprived urban areas. They may also be a dead end in the sense that there is not likely to be same level of dedicated investment in urban areas over the next few years – the language of co-ordination and partnership is usually accompanied by the proviso that there is no more money. Policy makers and politicians profess to believe that 'working together' will in itself improve things. In this context, it is, perhaps, worth reminding ourselves of the warning made by Higgins *et al.* (1983, p. 62) who point out that 'Co-ordination is an administrative opiate' which is frequently used to mask the lack of any focused action strategy.

Although all this could be seen as in sharp contradistinction to the programme of Urban Development Corporations, it is important to acknowledge some continuities, and maybe even to recognise that the experience of the Urban Development Corporations pointed in similar directions. They began to highlight strategies of partnership and co-ordination, and they undermined approaches which saw the inner cities as more or less innocent victims of global forces. In that sense they helped to point towards the new and more complex urban policy world of the 1990s, which combines a belief in 'proactive' localities, positioning themselves effectively within a world of global place marketing, and a belief in the need for inner city residents to find ways of positioning themselves effectively within emergent (and ever changing) labour markets (see, e.g., Peck, 1997). The urban development corporations sought to shift inner city areas from welfare to work (making them productive), just as the new urban social policy seeks to shift inner city residents from welfare to work.

Further reading

Developments in urban policy and urban governance in Britain have been closely linked in recent years. This is reflected in a range of books which focus on a series of related debates. In *Citizens and Cities. Urban Policy in the 1990s*, Hill (1994) uses developments in urban policy to explore the changing experience of urban citizenship in Britain since the 1970s. *Managing Cities. The New Urban Context* is edited by Patsy Healey, Stuart Cameron, Simin Davoudi, Stephen Graham and Ali-Madami Pour (1995). Although it does not directly consider the experience of the Urban Development Corporations, its chapters highlight some of the important directions of change in the management of urban areas. An earlier – but still highly relevant – assessment of the complex politics of urban policy and urban change can be found in Brindley, Rydin and Stoker (1989). Atkinson and Moon (1994) provide a comprehensive overview of the development of urban policy in Britain since the 1960s.

References

Agnew, J. and Duncan, J.S. (eds) (1989) *The Power of Place: Bringing Together Geographical and Sociological Imaginations*, Boston: Unwin Hyman

Alden, J., Bathy, M., Bathy, S., and Longley, P., (1988) An Economic and Social Profile of the Cardiff Bay Area *Cambria: a Welsh Journal of Geography*, 15, pp. 61–87

Alden, J. and Essex, S. (1999) The Cardiff Metropolitan Region, in Roberts, P., Thomas, K. and Williams, G. (eds), *Metropolitan Planning in Britain*, London: Jessica Kingsley

Allen, J., Massey, D. and Cochrane, A. with Charlesworth, J., Court, G., Henry, N. and Sarre, P. (1998) *Re-thinking the Region*, London: Routledge

Amin, A. and Robins, K. (1991) The re-emergence of regional economies? The mythical geography of flexible accumulation, *Environment and Planning D: Society and Space*, 8, 1, pp. 7–34

Amin, A. and Thrift N. (eds) (1994) *Globalisation, Institutions and Regional Development in Europe*, Oxford: Oxford University Press

Amin, A. and Thrift, N. (1995) Globalisation, institutional 'thickness' and the local economy, in Healey, P., Cameron, S., Davoudi, S., Graham, S. and Madani-Pour, A. (eds) *Managing Cities: The New Urban Context*, Chichester: John Wiley, pp. 91–108

Anderson, J. (1991) The new right, enterprise zones and urban development corporations, *International Journal of Urban and Regional Research*, 14, 3, pp. 468–489

Association of Island Communities (AIC) (1997) *Outstanding Regeneration Needs on the Isle of Dogs*, London: AIC

Association of Metropolitan Authorities (1986) *Programme for Partnership: an Urban Policy Statement*, London: AMA

Atkinson, M. and Coleman, W. (1992) Policy networks, policy communities, and the problems of governance, *Governance*, 5, 2, pp. 154–180

Atkinson, R. and Moon, G. (1994) *Urban Policy in Britain. The City, the State and the Market*, Basingstoke and London: Macmillan

Audit Commission (1989) *The Urban Regeneration Experience: Observations from Local Value for Money Audits*, London: HMSO

Baber, P. (1997) 'Closed doors are thrown open', *Planning*, 13 June, p. 10

Bailey, N., with Barker, A. and MacDonald, K. (1995) *Partnership Agencies in British Urban Policy*, London: UCL Press

Bassett, K. (1995) Partnerships, Business Elites and Urban Politics: New Forms of Governance in an English City? *Urban Studies*, 33, pp. 539–555

Bates, T. (1990) Building hopes and workplaces. Special issue on Business Opportunities on Merseyside, *The Daily Telegraph*, 21 March, p. 40

Batley, R. (1989) London docklands: an analysis of power relations between UDCs and local government, *Public Administration*, 67, 2, pp. 167–187

Benington, J. (1986) Local economic strategies: paradigms for a planned economy? *Local Economy*, 1, 1, pp. 7–24

Bentley, J. (1997) *East of the City*, London: Paul Chapman Publishing

Beynon, H., Elson, D., Howell, D., Peck, J. and Shaw, L. (1993) *The remaking of economy and society: Manchester, Salford and Trafford 1945–1992*. Manchester International Centre for Labour Studies Working Paper 1. Manchester: International Centre for Labour Studies

Beynon, H., Hudson R. and Sadler, D. (1994) *A Place called Teesside: A locality in a global economy*. Edinburgh: Edinburgh University Press

Bird, J., Curtis, B., Putnam, T., Robertson, G. and Tickner, L. (1993) (eds) *Mapping the Futures:*

Local Cultures, Global Change, London: Routledge

Black Training and Enterprise Group (1994), *Who Benefits from SRB?* London: BTEG

Blackman, T. (1995) *Urban Policy in Practice*, London: Routledge

Blunkett, D. and Jackson, K. (1987) *Democracy in Crisis: The Town Halls Respond*, London: The Hogarth Press

Board of Trade (1963) *The North East: A programme for regional development and growth.* Cmnd 2206. London: HMSO.

Boddy, M., Lovering, J. and Bassett, K. (1986) *Sunbelt City? A Study of Economic Change in Britain's M4 Growth Corridor*, Oxford: Oxford University Press

Bond, R., 1998, Down on the docks, *Surveyor*, March, pp. 13–14

Boyle, R. (1988) Private sector urban regeneration: the Scottish experience, in Parkinson, M., Foley, B. and Judd, D. (eds), *Regenerating the Cities: the UK crisis and the US experience,* Manchester: *Manchester University Press*, 74–93

Bradford, M. and Steward, A. (1988) Inner city refurbishment: an evaluation of private-public partnership schemes. *SPA Working Paper 2.* Manchester: School of Geography

Briggs, A. (1968) *Victorian Cities*, London: Penguin

Brindley, T., Rydin, Y. and Stoker, G. (1989) *Remaking Planning. The Politics of Urban Change in the Thatcher Years*, London: Unwin Hyman

Bristol Development Corporation (1989) *Vision for Bristol*, Bristol: Bristol Development Corporation

Brownill, S., (1990) *Developing London's Docklands: Another Great Planning Disaster?* London: Paul Chapman Publishing

Brownill, S (1993a) *Developing London's Docklands: Another Great Planning Disaster?* 2nd edition, London: Paul Chapman Publishing

Brownill, S. (1993b) The Docklands experience; locality and community in London, in Imrie, R. and Thomas, H. (eds) *British Urban Policy and the Development Corporations*, London: Paul Chapman Publishing, pp. 41–57

Brownill, S. (1998) From exclusion to partnership? The LDDC and community consultation, paper given to *Assessing the Legacies of the LDDC Conference*, University of East London, available from the author, School of Planning, Oxford Brookes University, Headington, Oxford

Brownill, S., Razzaque, K., Stirling, T. and Thomas, H. (1996) Local governance and the racialisation of urban policy in the UK: the case of UDCs, *Urban Studies* 33, 8, pp. 1337–1356

Brownill, S., Razzaque, K., Stirling, T. and Thomas, H., (1997) Race Equality and Local Governance, *ESRC Project Paper Number 3*, Department of City and Regional Planning, Cardiff University, PO Box 906, Cardiff, CF1 3EU

Burton, P. (1986) Planning theory and public policy: an analysis of policies for London's docklands, *unpublished PhD thesis*, School for Advanced Urban Studies, University of Bristol, Bristol

Burton, P. (1997) Urban policy and the myth of progress, *Policy and Politics*, 25, 4, pp. 421–437

Byrne, D. (1987) What is the point of a UDC for Tyne and Wear? in *Northern Economic Review*, No. 15, Summer pp. 63–73

Byrne, D. (1989), *Beyond the Inner City*, Milton Keynes: Open University Press

Byrne, D. (1992) The city, in Cloke, P. (ed.), *Policy and Change in Thatcher's Britain*, Oxford: Pergamon, pp. 247–268

Byrne, D. (1997) National social policy in the United Kingdom, in Pacione, M. (ed.) *Britain's Cities*, London: Routledge, pp. 108–127

Caborn, Richard (1997), *Regeneration Programmes: the Way Forward*, London: DETR

Cardiff Bay Community Trust (1998) *Business Plan*, Cardiff: CBCT

Cardiff Bay Development Corporation (1988) *Cardiff Bay Regeneration Strategy – The Summary*, Cardiff: CBDC

Cardiff Bay Development Corporation (1996a) *Sustaining Success*, Cardiff: CBDC

Cardiff Bay Development Corporation (1996b) *Corporate Plan 8*, Cardiff: CBDC

Cardiff Bay Development Corporation (1997a) *Corporate Plan 9*, Cardiff: CBDC

Cardiff Bay Development Corporation (1997b) *Annual Report 1996/97*, Cardiff: CBDC

Centre for Local Economic Strategies (1990a) Building a people's Europe, *Local Work: Monthly Bulletin of the Centre for Local Economic Strategies*, 21, CLES, Manchester

Centre for Local Economic Strategies (1990b) *Inner City Regeneration. A Local Authority Perspective.* First Year Report of the CLES Monitoring Project on Urban development Corporations. Manchester: CLES

Centre for Local Economic Strategies (1992) *Social Regeneration – Directions for Urban Policy in the 1990s*, Second report of the CLES Monitoring Project on Urban Development

Corporations, Manchester: CLES
Central Manchester Development Corporation, (1990) *Development strategy for Central Manchester*, Manchester: CMDC
Central Manchester Development Corporation (1996) *1988–1996: Eight years of achievement*, Manchester: CMDC
Central Manchester Development Corporation (no date) *Housing report*, Manchester: CMDC
City 20.20, (1997) *Final Report of the Inquiry into Urban Policy*, Keith Vaz MP, House of Commons, Westminster
Clarke, J. and Newman, J. (1997) *The Managerial State*, London: Sage
Clavel, P. and Kleniewski, N. (1990) Space for progressive local policy: examples from the United States and the United Kingdom, in Logan, J. and Swanstrom, T. (eds), *Beyond the City Limits*, New York: Temple University Press, pp. 199–236
Cochrane, A. (1983) 'Local Economic policies: trying to drain an ocean with a teaspoon' in Anderson, J., Duncan, S. and Hudson, R. (eds) *Redundant Spaces in Cities and Regions*, London: Academic Press, pp. 285–311
Cochrane, A. (1993) *Whatever happened to Local Government?* Buckingham: Open University Press
Cochrane, A., Peck, J. and Tickell, A. (1996) Manchester plays games: exploring the local politics of globalization, *Urban Studies*, 33, 8, pp. 1319–1336
Cohen, P. (1996) Out of the melting pot into the fire next time, in Westwood, S. and Williams, C. (eds) *Imagining Cities*, London: Routledge, pp. 73–85
Cohen, P. (1998) The road to Beckton pier? in *Rising East*, 1, 3, pp. 24–51
Colenutt, B., interview April 1988
Colls, R. and Lancater, W. (eds) (1992) *Geordies – Roots of Regionalism*, Edinburgh: Edinburgh University Press
Commission for Racial Equality (1991) *Employers in Cardiff*, London: CRE
Commission of the European Communities (1995) *Merseyside Single Programming Document, 1994–99*, Brussels: European Commission
Cooke, P. (1980) Capital relations and state dependency: an analysis of urban development policy in Cardiff, in Rees, G. and Rees, T. (eds) *Poverty and Social Inequality in Wales*, London: Croom Helm, pp. 206–229
Cooke, P. (1987) Clinical inference and geographic theory, *Antipode*, 19, 3, pp. 69–78
Cooke, P. (1989) Locality – theory and the poverty of 'spatial variation', *Antipode*, 21, 3, pp. 261–273
Coopers and Lybrand (1987) *Lower Don Valley: Final Report*, Sheffield: Sheffield Economic Regeneration Committee
Coulson, A. (1989) *Strategy and Impact of the Teesside Development Corporation*, Birmingham: University of Birmingham, Institute of Local Government Studies
Coulson, A. (1990) Flagships and flaws: assessing the UDC decade, *Town and Country Planning*, 59, 11, pp. 299–302
Coulson, A. (1993) Development Corporations, local authorities and patronage in urban policy, in Imrie, R. and Thomas, H. (eds) *British Urban Policy and the Urban Development Corporations*, 1st edition, London: Paul Chapman Publishing, pp. 27–38
Cox, K.R. (1998) Locality and community: some conceptual issues, *European Planning Studies*, 6, 1, pp. 17–30
Crick, M. (1986), *The March of Militant*, London: Faber and Faber
Crickhowell, N. (1997) *The Opera House Lottery*, Cardiff: University of Wales Press
Crocker, S. (1997) Links of Steel ? Business change in an industrial district; the Lower Don Valley in Sheffield. *Paper presented to XI AESOP Congress*, University of Nijmegen, The Netherlands, 28–31 May
Cullingworth, J. and Nadin, V. (1997) *Town and Country Planning in the UK*, 12th edn, London: Routledge
Dabinett, G. (1991) Local policies towards industrial change: the case of Sheffield's Lower Don Valley, *Planning Practice and Research* 6, 1, pp. 13–18
Dabinett, G. (1995) Economic Regeneration in Sheffield, in Turner, R. (ed) *The British Economy in Transition*, London: Routledge, pp. 218–239
Dabinett, G. and Lawless P. (1998) Sheffield : a city in search of a role and future, in Beazley, M., Nevin, B. and Loftman, P. (eds) *Local Communities and Mega-project development*, Aldershot: Avebury
Dalby, S. (1990) Heseltine's vision of land regeneration takes shape, in *Financial Times Survey*,

Urban Development in the Thatcher Era, 30 October, pp. 6–7

Dalby, S. (1992) Investment picks up, in *Financial Times Survey*, Merseyside, 2 July, p. 38

Darwin, J. (1990) *The Enterprise Society: regional policy and national strategy*, CLES Research Study 5, Manchester: Centre for Local Economic Strategies

Daunton, M.J. (1977) *Coal Metropolis: Cardiff 1870–1914*, Leicester: Leicester University Press

Davis, H. and Stewart, J. (1993) *The Growth of Government by Appointment*, Luton: Local Government Management Board

Dawson, J. and Parkinson, M. (1990), Urban Development Corporations: the Merseyside experience 1981–1990; Physical regeneration, political accountability and economic challenge, *Centre for Urban Studies Working Paper 13*, Liverpool: University of Liverpool

Deakin, N. and Edwards, J. (1993) *The Enterprise Culture and the Inner City*, London: Routledge

Deas, I. and Robinson, F. (1994) Squeezing the inner cities: local responses to changing urban policy priorities in Tyne and Wear, *SPA Working Paper 25*, Manchester: School of Geography, Manchester

Department of the Environment (1977) *Policy for the Inner Cities*, White Paper, London: HMSO

Department of the Environment (1979) Inner city policy: statement by Michael Heseltine, Secretary of State for the Environment, *Press Notice Number 390*, DOE, London

Department of the Environment (1988) *Action for Cities*, London: HMSO

Department of the Enviroment (1989) *Observations by the Govt on the third report of the Employment Committee*, HC 83 pp. 88–89, London: HMSO

Department of the Environment (1990) *Public expenditure on the Urban Development Corporations*, London: DOE

Department of the Environment (1994) *Assessing the Impact of Urban Policy*, London: HMSO

Department of the Environment (1996a) *City Challenge: Interim National Evaluation, Regeneration Research Summary No 9*, London: DoE

Department of the Environment (1996b), *Urban Development Corporations Accounts, 1994–1995*, London: HMSO

Department of the Environment (1996c) *Urban Development Corporations Accounts, 1995–1996*, London: HMSO

Department of the Environment (1997) *Annual Report*, London: HMSO

Department of the Environment, Transport and the Regions/Standing Advisory Committee on Trunk Road Assessment (1997) *Transport Investment, Transport Intensity and Economic Growth, Interim Report, December 1997*, London: HMSO

Department of the Environment, Transport and the Regions, (1997) *Regeneration Programmes – the way forward*, London: DETR

Department of Environment, Transport and the Regions (1998a) *Modernising Local Government*, London: Stationery Office

Department of the Environment, Transport and the Regions (1998b) *Local Democracy and Community Leadership*, Consultation Paper, London: DETR

DiGaetano, A. (1996) Urban governing alignments and realignments in comparative perspective: development politics in Boston and Bristol 1980–1985, *Urban Affairs Review*, 32, 6, pp. 844–70

DiGaetano, A. and Klemanski, J. (1993) Urban regimes in comparative perspective; the politics of urban development in Bristol, *Urban Affairs Quarterly*, 29, 1, pp. 54–83

Dobson N. and Gratton C. (1996) From city of steel to city of sport : an evaluation of Sheffield's attempt to use sport as a vehicle for urban regeneration, in Liu, A-H and Botterill D. (eds) *Higher Degrees of Pleasure*, Cardiff: University of Wales Press

Docklands Forum (1984) *Minutes of monthly meetings*, London: Docklands Forum

Docklands Forum (1987) *Housing in Docklands*, London: Docklands Forum

Docklands Forum (1994) Minutes of monthly meetings, London: Docklands Forum

Docklands Forum (1996) *Minutes of monthly meeting*, October, London: Docklands Forum

Du Gay, P. (1996) Organising identity; entrepreneurial governance and public management, in Hall, S. and Du Gay, P. (eds) *Questions of Cultural Identity*, London: Sage, pp. 151–169

Duncan, S. and Goodwin, M. (1988) *The Local State and Uneven Development*, London: Polity

Duncan, S. and Savage, M. (1989) Space, scale and locality, *Antipode*, 21, 3, pp. 179–206

Duncan, S. and Savage, M. (1991) New perspectives on the locality debate, *Environment and Planning A*, 23, 2, pp. 155–164

Duncan, W. (1888) *Industrial Rivers of the United Kingdom*, London

Eade, J. (1998) Global processes and customised landscapes, *Rising East* 1, 3, pp. 52–73

ECOTEC (1988) *Central Manchester Development Corporation: development strategy*, Birmingham: ECOTEC Research and Consulting Limited

Edgell, S. and Duke, V. (1991) *A Measure of Thatcherism*, London: Harper Collins

Edwards, B. (1992) *London Docklands: Urban Design in an Age of Deregulation*, London: Butterworth

Edwards, J. (1995) Social policy and the city, *Urban Studies*, 32, 4–5, pp. 695–712

Edwards, J. (1997) Urban policy: the victory of form over substance? *Urban Studies*, 34, 4–5, pp. 825–843

Edwards, J. and Batley, R. (1978) *The Politics of Positive Discrimination. An Evaluation of the Urban Programme 1967–77*, London: Tavistock

Ellin, N. (1996) *Postmodern Urbanism*, Oxford: Blackwell

ERM Economics (1997a) *Merseyside Development Corporation: Assessment of Housing and Population Outputs (1981–March 1996)*, London: ERM Economics

ERM Economics (1997b) *Merseyside Development Corporation: Assessment of Economic Outputs (1981–March 1996)*, London: ERM Economics

Evans, E. and Thomas, H. (eds) (1988) *Cardiff Capital Development*, Cardiff: Cardiff City Council

Evans, C., Dodsworth, S. and Barnett, J. (1984) *Below the Bridge*, Cardiff: National Museum of Wales

Fainstein, S., Gordon, I. and Harloe, M. (eds) (1992) *Divided Cities: New York and London in the Contemporary World*, Oxford: Blackwell

Financial Times (1986) Special Survey on 'Merseyside', 20 November

Financial Times (1989) Supplement on 'Merseyside in the Thatcher Decade', 19 October

Financial Times (1994) Special Survey on 'Merseyside', 14 July

Financial Times (1997) Special Survey on 'Merseyside', 3 April

Financial Times (1998) *Exit from London's docklands*, 19 March, London

Foley, P. (1991) The impact of major events: a case study of the World Student Games and Sheffield, *Environment and Planning C*, 9, pp. 65–78

Foord, J., Robinson, F. and Sadler, D. (1985) *The Quiet Revolution: Social and Economic Change on Teesside 1965–1985*, Report prepared for BBC North East, Newcastle: BBC

Geddes, M. (1997) *Partnerships against poverty and exclusion? Local regeneration strategies and excluded communities in the UK*, Bristol: The Polity Press

Goldsmith, M., (1992) Local Government, *Urban Studies*, 29, 3/4, pp. 393–410

Goodwin, M. (1991) Replacing a surplus population: the policies of the London Docklands Development Corporation, in Allen, J. and Hamnett, C. (eds), *Housing and Labour Markets: Building the Connections*, London: Unwin Hyman

Gosling, P. (1992) Feature on Urban Regeneration, *The Guardian*, 23 July, p. 18

Greer, A. and Hoggett, P. (1996) Quangos and local governance, in Pratchett, L. and Wilson, D. (eds) *Local Democracy and Local Government*, London: Macmillan

Gregory, D. and Urry, J. (eds) (1985) *Social Relations and Spatial Structures*, London: Macmillan

Griffiths, S. (1998) *Manchester poverty profile*, Manchester: Manchester City Council and Central Manchester Health Authority

Grimley J. R. Eve (1995) *Manchester office investment review, Autumn*. Manchester: Grimley, J. R. Eve

The Guardian (1998) Hit squads' for sink estates, 24 January

Gurr, T. and King, D. (1987) *The State and the City*, London: Macmillan

Gyford, J. (1985) *The politics of local socialism*, London: Allen and Unwin

Hague, C. and Thomas, H. (1997) Capital city planning: Cardiff and Edinburgh compared, in Macdonald, R. and Thomas, H. (eds) *Nationality and Planning in Scotland and Wales*, Cardiff: University of Wales Press, pp. 133–158

Hain, P. (1998), *South Wales Echo*, 9 March, Cardiff

Hall, D. (1997) Bridges to Better Government, *Local Government Chronicle*, 28 November

Hall, T. (1998) *Urban Geography*, London: Routledge

Hall, T. and Hubbard, P. (1996) The entrepreneurial city: new urban politics, new urban geographies?, *Progress in Human Geography*, 20, 2, pp. 153–174

Hambleton, R. (1995) The Clinton Policy for Cities: a transatlantic assessment, *Planning Practice and Research* 10, 3/4, pp. 359–378

Hambleton, R. and Thomas, H. (eds) (1995) *Urban Policy Evaluation*, London: Paul Chapman Publishing

Hamilton Fazey, I. (1989) Labour and business at peace in Liverpool, *Financial Times*, 19 June

Hansard (1981) Merseyside Development Corporation, *Parliamentary Debates, Sixth Series*, Volume 1, Session 1980–81, 19 March, cols 511–530

Hansard (1987) Urban Development Corporations (Financial Limits Bill), *Parliamentary Debates, Sixth Series,* Volume 121, Session 1987–88, 4 November 1987, col. 943

Hansard (1989) Parliamentary Answer, *UDC Spending on Community projects,* May, col. 265

Hansard (1992a), *Sixth Series,* Volume 210, Number 38, Session 1991–92, 29 June, Written Answers, col. 367

Hansard (1992b), *Sixth Series,* Volume 210, Number 35, Session 1991–92, 24 June, Written Answers, col. 196

Hansard (1992c), *Sixth Series,* Volume 210, Number 47, Session 1991–92, 9 July, Written Answers, cols. 307–310

Hansard (1992d), *Sixth Series,* Volume 210, Number 48, Session 1991–92, 10 July, Written Answers col. 390

Harding, A. (1991) The rise of urban growth coalitions, UK-style, *Environment and Planning C: Government and Policy,* 9, pp. 295–317

Harloe, M. and Fainstein, S. (1992), Introduction, in Fainstein, S., Gordon, I. and Harloe, M. (eds) *Divided Cities: New York and London in the Contemporary World,* Oxford: Blackwell, pp. 1–28

Hart, S. (1998) Nothing Can Stop Us Now, *Big Issue,* 13 April

Harvey, D. (1987) Three myths in search of a reality in urban studies, *Environment and Planning D: Society and Space,* 5, pp. 367–376

Harvey, D. (1989a) *The Condition of Post-modernity,* Oxford: Basil Blackwell

Harvey, D. (1989b) *The Urban Experience,* Oxford: Basil Blackwell

Harvey, D. (1989c), Transformation in urban governance in late capitalism, *Geografiska Annaler,* 71 (B), pp. 3–17

Harvey, D. (1993) From space to place and back again: reflections on the condition of postmodernity, in Bird, J., Curtis, B., Putnam, T., Robertson, G. and Tickner, L. (eds) *Mapping the Futures: Local Cultures, Global Change,* London: Routledge

Harvey, D. (1997a), Contested cities: social process and spatial form, in Jewson, N. and MacGregor, S. (eds), *Transforming Cities: Contested Governance and New Spatial Divisions,* London: Routledge, pp. 19–27

Harvey, D. (1997b) *Justice, Nature and the Geography of Difference,* Oxford: Blackwell

Hayes, M. (1987) *Urban Development Corporations: the Liverpool Experience,* Liverpool: Liverpool City Council

Headicar, M. (1995) UDCs and Regeneration: Early Lessons Learnt. *Discussion Paper*

Healey, P. (1991) Urban regeneration and the development industry, *Regional Studies,* 25, 2, pp. 97–110

Healey, P. (1992) Urban Policy and Property Development, *paper presented at the 6th AESOP Conference,* 3–6 June, Stockholm, Sweden

Healey, P. and Barrett, S. (1990) Structure and agency in land and property development processes: some ideas for research, *Urban Studies;* 27, 1, pp. 89–104

Healey, P., Cameron, S., Davoudi, S., Graham, S. and Madandi-Pour, A. (eds) (1995) *Managing Cities: the new Urban Context,* Chichester: John Wiley and Sons

Hey, D., Olive, M. and Liddament, M. (1997) *Forging the Valley,* Sheffield: Sheffield Academic Press

Higgins, J., Deakin, N., Edwards, J. and Wicks, M. (1983) *Government and Urban Poverty. Inside the Policy-making Process,* Oxford: Basil Blackwell

Hill, C. (1996) *Olympic Politics: Athens to Atlanta 1896–1996.* Manchester: Manchester University Press

Hill, D. (1994), *Citizens and Cities: Urban Policy in the 1990s,* Hemel Hempstead: Harvester, Wheatsheaf

House of Commons Public Accounts Committee (1988) *Twentieth report: Urban Development Corporations,* London: HMSO

House of Commons Select Committee on Employment (1989) *Third Report – The Employment Effects of UDCs,* HC 327 I and II, London: HMSO

Hudson, R. (1990) Trying to revive an infant Hercules: the rise and fall of local authority modernisation policies on Teesside, in Harloe, M., Pickvance, C. and Urry, J. (eds) *Place, Policy and Politics: Do Localities Matter?* London: Unwin Hyman

Imrie, R. (1993) Interview conducted with planning official in Cardiff Bay Development Corporation, 23 May

Imrie, R. (1996) Transforming the social relations of research production in Urban Policy evaluation, *Environment and Planning A,* 28, 8, pp. 1445–1464

Imrie, R. (1997) National economic policy in the United Kingdom, in Pacione, M. (ed) *Britain's Cities: Geographies of Division in Urban Britain*, Harlow: Longman, pp. 88–107

Imrie, R. and Raco, M., (1999) How new is the new local governance? Lessons from the United Kingdom, *Transactions of the Institute of British Geographers: New Series* 24, 1, (forthcoming)

Imrie, R. and Thomas, H. (1992) The wrong side of the tracks: a case study of local economic regeneration in Britain. *Policy and Politics*, 20, 3, pp. 213–226

Imrie, R. and Thomas, H. (1993a) The new partnership: the local state and the property development industry, in Ball, R. and Pratt, A. (eds) *Industrial Property: Policy and Economic Development*, London: Routledge, pp. 129–151.

Imrie, R. and Thomas, H. (eds) (1993b) *British Urban Policy and the Urban Development Corporations*, 1st edn, London: Paul Chapman Publishing

Imrie, R. and Thomas, H. (1993c) Urban policy and the urban development corporations in Imrie, R. and Thomas, H. (eds) *British Urban Policy and the Urban Development Corporations.*, 1st edn, London: Paul Chapman Publishing, pp. 3–26

Imrie, R. and Thomas, H. (1995) Urban policy processes and the politics of urban regeneration. *International Journal of Urban and Regional Research*, 18, 4, pp. 479–494

Imrie, R. and Thomas, H. and Marshall, T. (1995), Business organizations, local dependence and the politics of Urban Renewal in Britain, *Urban Studies*, 32, 1, pp. 31–47

Jessop. B. (1997), The entrepreneurial city: re-imagining localities, redesigning economic governance, or restructuring capital?, in Jewson, N. and MacGregor, S. (eds), *Transforming Cities: Contested Governance and New Spatial Divisions*, London: Routledge, pp. 28–41

Jewson, N. and MacGregor, S. (1997a) Transforming cities: social exclusion and the reinvention of partnership, in Jewson, N. and MacGregor, S. (eds), *Transforming Cities: Contested Governance and New Spatial Divisions*, London: Routledge, pp. 1–18

Jewson, N. and MacGregor, S. (1997b) *Transforming Cities: Contested Governance and New Spatial Divisions*, London: Routledge

Johnson, R.W. 1925 *The Making of the River Tyne*, Newcastle: Andrew Reid

Jones, D. (1989) *A Catalyst for Regeneration*, Newcastle: TWDC, p. 3

Kapp, F.W. (1978) *The Social Costs of Business Enterprise*, Nottingham: Spokesman Books

Kirkham, S. (1990) *Sheffield Development Corporation and the Lower Don Valley. A Study for the Joint Initiative for Social and Economic Research*, Sheffield : Sheffield Business School

Kitchen, T. (1997) *People, Politics, Policies and Plans: the City Planning Process in Contemporary Britain*, London: Paul Chapman Publishing

Knowsley Metropolitan Borough Council (1989) *Urban Renewal Programme Submission 1989/90–1992/93*, Knowsley Metropolitan Borough Council

KPMG (1996) *Merseyside Economic Assessment*, KPMG in association with Cambridge Econometrics and the University of Warwick Institute of Employment Research, Liverpool: KPMG

KPMG (1997a) *Merseyside Development Corporation: Post Implementation Project Appraisal, Albert Dock*, Liverpool: KPMG

KPMG (1997b) *Merseyside Development Corporation: Post Implementation Project Appraisal, Vauxhall Housing Development*, Liverpool: KPMG

Lane, T. (1997), *Liverpool: City of the Sea*, Liverpool: Liverpool University Press

Law, C. and Dundon-Smith, D. (1995) Metrolink impact study, *Department of Geography Working Paper 13*, Salford: Salford University

Lawless, P. (1989) *Britain's Inner Cities*, London: Paul Chapman Publishing

Lawless, P. (1990) Regeneration in Sheffield: from radical intervention to partnership, in Parkinson, M. and Judd, D. (eds) *Leadership and Urban Regeneration*, London: Sage, pp. 133–151

Lawless, P. (1991) Urban policy in the Thatcher decade: English inner-city policy, 1979–90, *Environment and Planning C: Government and Policy*, 9, pp. 15–30

Lawless P. (1995) Inner city and suburban labour markets in a major English conurbation : process and policy implications, *Urban Studies* 32, 7, pp. 1097–1125

Lawless, P. (1996) The inner cities. Towards a new agenda, *Town Planning Review*, 67, 1, pp. 21–43

Lewis, N. (1992) *Inner City Regeneration. The Demise of Regional and Local Government*, Buckingham: Open University Press

London Docklands Development Corporation, [LDDC] (1987), *Annual Report and Accounts*, London: LDDC

LDDC, (1985) *Annual Report and Accounts*, London: LDDC

LDDC (1990) *Annual Report and Accounts*, London: LDDC

LDDC (1991) *Annual Report and Accounts*, London: LDDC
LDDC (1992) *Annual Report and Accounts*, London: LDDC
LDDC (1995) *Annual Report and Accounts*, London: LDDC
LDDC (1997a) *Attitudes of Local Residents*, London, LDDC
LDDC (1997b) *Key Facts and Figures*, London: LDDC
LDDC (1997c) *Starting From Scratch*, London: LDDC
LDDC (1998a) *The Battle For Docklands*, London: LDDC
LDDC (1998b) *The 1997 Docklands Employment Census*, London: LDDC
Liverpool City Council (1991) *Economic Development Plan 1991/92*, Liverpool: Liverpool City Council
Liverpool City Council (1993) *Urban Policy: Looking to the Future*, Liverpool: Central Policy Unit, Liverpool City Council
Liverpool City Council (1997) *Draft Economic Development Plan 1996/97*, Liverpool: Liverpool City Council
Liverpool Echo, (1996), *Rebirth of Merseyside: The Future is Bright,* 8 November
Lloyd, P.E. and Meegan, R.A. (1996) Contested governance: European exposure in the English regions, *European Planning Studies*, 4, 1, pp. 75–97
Local Government Association, (1998) *The New Commitment to Regeneration*, London: LGA
LGPLA, (1980) *Local Government, Planning and Land Act,* HMSO: London
Loftman, P. (1998) Urban Partnerships, *Urban Environment Today*, February
Loftman, P and Beazley, M. (1998) *Race and Regeneration*, London: LGIU
Loftman, P. and Nevin, B. (1996) Going for growth: prestige projects in three British cities, *Urban Studies*, 33, 6, pp. 991–1019
Logan, J. and Swanstrom, T. (eds) (1990) *Beyond the City Limits*, New York: Temple University Press
Loney, M. (1983) *Community against Government. The British Community Development Project 1968–78 – a Study of Government Incompetence*, London: Heinemann
Lovatt, A. (1996) The ecstasy of urban regeneration: regulation of the night-time economy in the transition to a post-Fordist city, in O'Connor, J. and Wynne, D. (eds) *From the Margins to the Centre: Cultural Production and Consumption in the Post-Industrial City*, Aldershot: Arena, pp. 141–168
Lovering, J. (1997) Global restructuring and local impact, in Pacione, M. (ed), *Britain's Cities: Geographies of Division in Urban Britain*, Harlow: Longman, pp. 63–87
Lowndes, V., Nanton, P., McCabe, A. and Skelcher, C. (1997) Networks, Partnerships and Urban Regeneration, *Local Economy*, 12, 4, pp. 333–342
Mabbott, J. (1992), *Local Authority Funding for Voluntary Organizations: Report and Recommendations*, London: National Council for Voluntary Organisations
Macfarlane R. (1993) *Community Involvement In City Challenge*, London: National Council for Voluntary Organisations
Mackintosh M. (1992) Partnerships : issues of policy and negotiation, *Local Economy* 7, 3, pp. 210–224
Malpass P. (1994) Policy making and local government: how Bristol failed to secure City Challenge funding (twice), *Policy and Politics*, 22, 4, pp. 301–12
Manchester City Council (1990) *Office Development in Manchester*, Manchester: Manchester City Council
Manchester City Council (1995) *Office Development in Manchester*, Manchester: Manchester City Council
Massey, D. (1985) New directions in space, in Gregory, D. and Urry, J. (eds) *Social Relations and Spatial Structures*, London: Macmillan, pp. 9–19
Massey, D. (1992) A place called home?, *New Formations*, 17, Summer, pp. 3–15
Massey, D. (1993a) Power-geometry and a progressive sense of place, in Bird, J., Curtis, B., Putnam, T., Robertson, G. and Tickner, L. (eds) *Mapping the Futures: Local Cultures, Global Change*, London: Routledge, pp. 59–69
Massey, D. (1993b) Politics and space/time, in Keith, M. and Pile, S. (eds) *Place and the Politics of Identity*, London: Routledge, pp. 141–161
Massey, D. (1994) A Global Sense of Place in *Society, Place, and Gender*, Minnesota: University of Minnesota Press and London: Polity Press
Massey, D. (1995) *Spatial Divisions of Labour: Social Structures and the Geography of Production*, 2nd edn, London: Macmillan
Mayer, M. (1989) Local politics: from administration to management, *paper presented at the*

Cardiff symposium on regulation, innovation, and spatial development, University of Cardiff, 13–15 September

Mayo, M. (1996) Partnerships for regeneration and community development: some opportunities, challenges and constraints, *Critical Social Policy*, 17, 3, pp. 3–26

McAllister, D., (1980) *Evaluation in Environmental Planning*, London: MIT Press

Meegan, R. (1989) Paradise postponed; the growth and decline of Merseyside's outer estates, in Cooke, P. (ed), *Localities: the Changing Face of Urban Britain*, London: Unwin Hyman, pp 198–234

Meegan, R. (1990) Merseyside in crisis and in conflict, in Harloe, M., Pickvance, C. and Urry, J. (eds), *Place, Policy and Politics: Do Localities Matter?*, London: Unwin Hyman

Meegan, R. (1995) Local Worlds in Allen, J. and Massey, D. (eds) *Geographical Worlds*, Oxford: Oxford University Press, pp. 53–104

Mellor, R. (1997) Cool times for a changing city, in Jewson, N. and MacGregor, S. (eds) *Transforming Cities, Contested Governance and New Spatial Divisions*, London: Routledge, pp. 56–72

Merseyside Development Corporation [MDC] (1981) *Initial Development Strategy Merseyside Development Corporation: Proposals for the Regeneration of Its Area*, Liverpool: Merseyside Development Corporation

MDC (1989) *Annual Report and Financial Statements for the Year Ended 31 March 1989*, Liverpool: Merseyside Development Corporation

MDC (1992a) *Corporate Plan*, Liverpool: Merseyside Development Corporation

MDC (1992b) *Annual Report for the Year Ending 31st March 1992*, Liverpool: Merseyside Development Corporation MDC (1994) *Financial Statements for the Year Ended 31 March 1994*, Liverpool: Merseyside Development Corporation

MDC (1997a) *All Our Tomorrows: Merseyside Development Corporation, 1981–1998*, Liverpool: Merseyside Development Corporation

MDC (1997b) *Completion, Exit and Succession Plan 1997/8: Final Draft*, Liverpool: Merseyside Development Corporation

MDC (1997c) *Merseyside Development Corporation Regeneration Statement: Draft for Consultation*, Liverpool: Merseyside Development Corporation

MDC (1997d) *Annual Report, 1996/97*, Liverpool: Merseyside Development Corporation

Mess, H.A. (1928) *Industrial Tyneside*, London: Ernest Bess

Milburn, G.E. and Miller, S.T. (1988) *Sunderland: River, Town and People*, Sunderland: Sunderland Libraries

Moore, C. and Richardson, J. (1989) *Local partnership and the unemployment crisis in Britain*, London: Unwin Hyman

Moore, R. (1997) *Positive Action in Action: Equal Opportunities and Declining Opportunities on Merseyside*, Aldershot: Ashgate

National Audit Office (1988) *Department of the Environment: Urban Development Corporations*, London: HMSO

National Audit Office (1990) *Regenerating the Inner Cities*, London: HMSO

National Audit Office (1993) *The Achievements of the Second and Third Generation Urban Development Corporations*, London: HMSO

National Audit Office (1994) *Merseyside Development Corporation: Grand Regatta Colombus and fanfare for a New World*, London: HMSO

National Audit Office (1995) *LDDC: The Limehouse Link*, HC468 94/95, London: HMSO

National Audit Office (1997) *Wind up of Leeds and Bristol UDCs*, HC 292 96/97, London: HMSO

National Council for Voluntary Organisations (1994) *Community Involvement in City Challenge – A Policy Report*, London: NCVO

Nevin, B. and Shiner, P. (1995) Community regeneration and empowerment: a new approach to partnership, *Local Economy*, 19, 4, pp. 308–322

Newman, P. and Thornley, A. (1996) *Urban Planning in Europe*, London: Routledge

North, G. (1975) *Teesside's Economic Heritage*. Middlesbrough: Cleveland County Council

North Tyneside Borough Council, (1988), *Observations on Planning Proposals*, Newcastle-upon-Tyne: NTBC.

O'Toole M. (1996), *Regulation Theory and the British State: The Case of the Urban Development Corporation*, Aldershot: Avebury

Oatley, N. (1993) Realizing the potential of urban policy: the case of the Bristol Development Corporation, in Imrie, R. and Thomas, H. (eds) *British Urban Policy and the Urban Devel-*

opment Corporations, 1st edition London: Paul Chapman Publishing, pp. 136–153

Oatley, N. (1995) Competitive urban policy and the regeneration game, *Town Planning Review*, 66, 1, pp. 1–14

Oatley, N. (ed.) (1998) *Cities, Economic Competition and Urban Policy*, London: Paul Chapman Publishing

Oatley, N. and Lambert, C. (1995) Evaluating competitive urban policy: the City Challenge initiative, in Hambleton, R. and Thomas, H. (eds) *Urban Policy Evaluation. Challenge and Change*, London: Paul Chapman Publishing, pp. 141–57

Oatley, N. and Lambert, C. (1998) Catalyst for change: the City Challenge Initiative, in Oatley, N. (ed), *Cities, Economic Competition and Urban Policy*, London: Paul Chapman Publishing, pp. 109–126

Observer, (1994), *Feature on London docklands*, 9 October, London.

Ogden, I. (ed) (1992) *London Docklands: The Challenge of Development*, Cambridge: Cambridge University Press

Osborne, D. and Gaebler, T. (1995) *Reinventing Government: How the Entrepreneurial Spirit is Transforming the Public Sector*, Reading, Massachusetts: Addison Wesley

Open University (1997) *Your Place or Mine?* Programme for D (Media, Culture, Representation) BBC Open University Programmes: Milton Keynes

PA Cambridge Economic Consultants (1991) *Central Manchester Development Corporation Office Study*, Manchester: CMDC

Pacione, M. (1990), What about people? A critical analysis of urban policy in the United Kingdom, *Geography*, 75, 3, no. 328, pp. 193–202

Pacione, M. (1997) Urban restructuring and the reproduction of inequality in Britain's cities: an overview, in Pacione, M. (ed) *Britain's Cities: Geographies of Division in Urban Britain*, London: Routledge, pp. 7–62

Parkinson, M. (1989), The cities under Mrs Thatcher; the centralisation and privatisation of power, *Centre for Urban Studies Working Paper 6*, Liverpool: University of Liverpool

Parkinson, M. (1990) Leadership and regeneration in Liverpool: confusion, confrontation or coalition?, in Judd, D. and Parkinson, M. (eds) *Leadership and Urban Regeneration: Cities in North America and Europe*, Newbury Park: Sage, pp. 241–257

Parkinson, M. and Evans, R. (1988) Urban regeneration and development corporations: Liverpool style, *Centre for Urban Studies Working Paper 2*, Liverpool: University of Liverpool

Parkinson, M. and Evans, R. (1989) Urban development corporations, *Centre for Urban Studies Working Paper 3*, Liverpool: University of Liverpool

Parkinson, M. and Wilks, S. (1985) The politics of inner city partnerships, in Goldsmith, M. (ed) *New Research in Central-Local Relations*, Aldershot: Gower, pp. 290–307

Peck, J. (1995) Moving and shaking: business elites, state localism and urban privatism, *Progress in Human Geography*, 19, 1, pp. 16–46

Peck, J. (1997) *Work-Place: The Social Regulation of Labor Markets*, New York: Guildford

Peck, J. and Emmerich, M. (1992) Recession, restructuring . . . and recession again: the transformation of the Greater Manchester labour market, *Manchester Geographer*, 13, pp. 19–46

Peck, J. and Tickell, A. (1994). Too many partners . . . the future of regeneration partnerships, *Local Economy* 9, 3, pp. 251–265

Peck, J. and Tickell, A. (1995) Business goes local: dissecting the 'business agenda' in Manchester, *International Journal of Urban and Regional Research* 19, 1, pp. 55-78

Peck J. and Tickell, A. (1997) Manchester's jobs gap, *Manchester Economy Group Working Paper 1*, Manchester: Manchester Economy Group, University of Manchester

Peel, M. (1997) A Model that Works, in *Financial Times Guide: Business in the Community, Financial Times*, pp. 24–25

Pickard, M., (1988), Editorial, *Docklands News*, April, London: LDDC.

Pickup, J. (1988) Cardiff Bay – the way forward, in Evans, E. and Thomas, H. (eds) *Cardiff Capital Development*, Cardiff: Cardiff City Council, p. 10

Pickvance, C.G. (1998) Locality, local social structure and growth politics: a comment on Cox's 'Locality and community: some conceptual issues', *European Planning Studies*, 6, 1, pp. 43–47

Pike, A. (1997) UDCs prepare to sign off with much show, *Financial Times Survey: Reporting Britain, Financial Times*, 18 December, p. 1

Property Guide (1998) Property around the city and Docklands 3, 1

Price Waterhouse, Storey Sons and Partners, Mott Hay and Anderson, and Brian Clouston and Partners (1989) *The formation of the Tyne and Wear Development Corporation*, London: Department of the Environment

Quilley, S. (1995) Economic transformation and local strategy in Manchester, *unpublished PhD dissertation*, Department of Sociology, University of Manchester

Raco, M. (1997) Business associations and the politics of urban renewal: the case of the Lower Don Valley, Sheffield, *Urban Studies* 34, 3, pp. 383–402

Randall, S. (1995) City Pride – from 'municipal socialism' to 'municipal capitalism', *Critical Social Policy* 43, 8, pp. 40–59

Redclift, M. (1987) *Sustainable Development: Exploring the Contradictions*, London: Routledge

Redwood, J. (1992) Interview on Radio 4, *The World This Weekend*, Sunday, August 16

Rees, G. and Lambert, J. (1981) Nationalism as legitimisation? Notes towards a political economy of regional development in South Wales, in Harloe, M. (ed) *New Perspectives in Urban Change and Conflict* London: Heinemann Educational, pp. 122–137

Regan, M. (1990) Waterfront tourist trap in Liverpool, *Financial Times*, 30 October

Richard Ellis (1996) *Central Manchester office survey*, February, London: Richard Ellis

Roberts, P. and Whitney, D. (1993) The new partnership: inter-agency cooperation and urban policy in Leeds, in Imrie, R. and Thomas, H. (eds) *British Urban Policy and the Urban Development Corporations*, 1st edn, London: Paul Chapman Publishing, pp. 154–172

Robinson, F. (1989) Urban Regeneration Policies in Britain in the Late 1980s. Who Benefits? *Discussion Paper No, 94*, Newcastle: CURDS, University of Newcastle

Robinson F. (1997) *The City Challenge Experience*. Newcastle: Newcastle City Challenge West End Partnership Ltd

Robinson, F. (1998) *Stockton City Challenge: An Evaluation of Processes and Achievements*, Stockton: Stockton City Challenge

Robinson, F. and Shaw, K. (1991) Urban regeneration and community development, *Local Economy*, 6, 1, pp. 61–72

Robinson, F., Shaw, K. and Lawrence, M. (1993) *More than Bricks and Mortar? Tyne and Wear and Teesside Development Corporations: A mid-term report*, Durham: University of Durham

Robson, B. (1988) *Those Inner Cities*, Oxford: Clarendon

Robson, B., Bradford, M., Deas, I., Hall, E., Harrison, E., Parkinson, M., Evans, R., Harding, A., Garside, P. and Robinson, F. (1994), *Assessing the impact of urban policy*, London: HMSO.

Robson, B., Bradford, M., Deas, I., Fielder, A. and Franklin, S. (1998a) Evaluation of UDCs at Bristol, Central Manchester and Leeds, unpublished report to Department of the Environment Transport and the Regions. Department of the Environment Transport and the Regions, London

Robson, B., Deas, I. and Bradford, M. (1998b) Beyond the boundaries: vacancy chains and the evaluation of Urban Development Corporations, *Environment and Planning A*, forthcoming

Rowley, G. (1994), The Cardiff Bay Development Corporation: urban regeneration, local economy and community, *Geoforum*, 25, 3 pp. 265–84

Royal Institution of Chartered Surveyors (1995) *The Private Finance Initiative, the Essential Guide*, London: RICS

Russell, H. (1996) *The Costco Employment Initiative*, European Institute for Urban Affairs, Liverpool: Liverpool John Moores University

Russell, H., *et al.*, (1996), *City Challenge: Interim National Evaluation*, London: HMSO

Rustin T. (ed) (1996) *Rising in the East: The Regeneration of East London*, London: Lawrence and Wishart

Sadler, D. (1990) The social foundations of planning and the power of capital: Teesside in historical context, in *Environment and Planning D: Society and Space*, 8, pp. 323–338

Sassen, S. (1991) *The Global Cities: London, New York, Tokyo*, Princeton: Princeton University Press

Savills (1997) *Quarterly Residential Bulletin, Summer 1997*, London: Savills

Seyd, P. (1990) Radical Sheffield: from socialism to entrepreneurialism, *Political Studies*, 38, 2, pp. 335–344

Shaw, K. (1990) The politics of public-private partnerships in Tyne and Wear, *Northern Economic Review*

Shaw, K (1995) Assessing the Performance of Urban Development Corporations: How reliable are the official government output measures? *Planning Practice and Research*, 10, 3/4, pp. 287–298

Shaw, K., Robinson, F. and Curran, M. (1996) Non-Elected Agencies and Quangocrats; A Profile of Board Membership in the North-East of England, *Regional Studies*, 30, 3, pp. 295–299

Sheffield City Council (1986) *Urban Programme – City of Sheffield 1986–89*, Sheffield: Sheffield City Council

Sheffield City Council (1990a) *The World Student Games Economic Impact Study Report: Part*

1, Department of Employment and Economic Development, Sheffield: Sheffield City Council

Sheffield City Council (1990b) *Sheffield 2000: Phase One*, Report of the Sheffield Economic Regeneration Committee, Sheffield: Sheffield City Council

Sheffield City Council (1991a) *Urban Programme – City of Sheffield 1991–94*, Chief Executive, Sheffield : Sheffield City Council

Sheffield City Council (1991b) *The Vision – Sheffield City Challenge 1991*, Sheffield: Sheffield City Council

Sheffield City Liaison Group (1995) *Sheffield Shaping the Future*, Sheffield: Sheffield City Liaison Group

Sheffield City Liaison Group (1996) *Sheffield Growing Together 1996*, Sheffield: Sheffield City Liaison Group

Sheffield Development Corporation (1989) *A Vision of the Lower Don Valley: A Planning Framework for Discussion*, Sheffield: Sheffield Development Corporation

Sheffield Development Corporation (1991) *Report and Accounts 1990–91*, Sheffield: Sheffield Development Corporation

Sheffield Development Corporation (1993) *City of Manufacturing Excellence*, Sheffield: Sheffield Development Corporation

Sheffield Development Corporation (1996) *Into the Home Straight – Annual report and accounts 1995/1996*, Sheffield: Sheffield Development Corporation

Sheffield Development Corporation (1997) *Regeneration Statement*, Sheffield: Sheffield Development Corporation

Sherwood, M. (1991) Racism and resistance: Cardiff in the 1930s and 1940s. *Llafur. Journal of Welsh Labour History*, 5, 4, pp. 51–70

Shields Gazette (1997) 14 November

Short, J. (1989) Yuppies, yuffies and the new urban order, *Transactions of the Institute of British Geographers: New Series*, 14, 1, pp. 173–188

Silke, E. (1997) Bristol broke rules and blundered on road, *Surveyor*, 6 March, p. 3

Simmons, M. and Warren, A. (1998) Redressing the East-West balance in LPAC's East London Development Focus, *Rising East*, 1, 3, pp. 74–86

Skelcher, C., Lowndes, V. and McCabe, A. (1996) *Community Networks in Urban Regeneration*, Bristol: Policy Press

Smith, N. (1987) Dangers of the empirical turn: some comments on the CURS initiative, *Antipode*, 19, 1, pp. 59–68

Snape, D. and Stewart, M. (1995) *Bristol and the West: The Continuing Partnership Challenge*, report from the School of Policy Studies, Bristol: University of Bristol and the Chamber of Commerce

Solesbury, W. (1990) Property development and urban regeneration, in Healey, P. and Nabarro, R. (eds), *Land and Property in a Changing Context*, Aldershot: Gower, pp. 186–194

Stephenson D, interviewed by Nick Oatley, December 1997

Stewart, J. (1997) *The Need for Community Governance, in A Framework for the Future*, London: Local Government Information Unit

Stewart, M. (1994) Between Whitehall and town hall: the realignment of urban policy in England, *Policy and Politics*, 22, 2, pp. 133–145

Stewart M. (1996a) Too Little, Too Late: The Politics of Local Complacency, *Journal of Urban Affairs*, 18, 2 pp. 119–137

Stewart, M. (1996b) Urban Policy in Thatcher's England, School for Advanced Urban Studies (now School for Policy Studies), *Working Paper No 90*, University of Bristol

Stockton Metropolitan Borough Council. (1997) *The Further Progress of the Teesside Development Corporation*, January 1996–December 1996, Stockton: SMBC

Stoker, G. (1989) Urban Development Corporations: a review, *Regional Studies*, 23, 2, pp. 159–173

Stoker, G. (1991) *The Politics of Local Government*, London: Macmillan

Strange, I. (1997) Directing the show? Business leaders, local partnership, and economic regeneration in Sheffield, *Environment and Planning C, Government and Policy*, 15, 1, pp. 1–17

The Sunday Sun, January 1998

Taylor, I., Evans, K. and Fraser, P. (1996) *A tale of two cities: global change, local feeling and everyday life in the north of England*, London: Routledge

Teesside Development Corporation (1997) *Annual Report and Financial Statements for the Year Ended 31 March 1997*, Middlesborough: TDC

Teesside Development Corporation (1998) *Regeneration Statement*, Middlesborough: TDC

Thake, S. and Staubach, R. (1993) *Investing in People. Rescuing Communities from the Margin*, York: Joseph Rowntree Foundation

Thames Gateway London Partnership (TGLP) (undated) *Economic Strategy; Thames Gateway in the 21st Century*, London (TGLP)

The Times (1992) Focus: Merseyside, 13 July, 23

Thomas, H. (1989) Cardiff-City Profile, *Cities* 6, 2, pp. 91–101

Thomas, H. (1992) Redevelopment in Cardiff Bay: State intervention and the securing of consent, *Contemporary Wales*, 5, pp. 81–98

Thomas, H. (1999) Spatial restructuring in the capital: struggles to shape Cardiff's built environment, in Fevre, R. and Thompson, D. (eds) *Nation, identity and social theory*, Cardiff: University of Wales Press (forthcoming)

Thomas, H. and Imrie, R. (1989) Urban redevelopment, compulsory purchase and the regeneration of local economies: the case of Cardiff Bay, *Planning Practice and Research*, 4, 3, pp. 18–27

Thomas, H. and Imrie, R. (1993b) Cardiff Bay and the Project of Modernisation in Imrie, R. and Thomas, H. (eds) *British Urban Policy and the Urban Development Corporations* 1st edn, London: Paul Chapman Publishing, pp. 74–88

Thomas, H. and Imrie, R. (1997) Urban Development Corporations, Urban Regimes and Local Governance in the United Kingdom, *Tijdschrift Voor Economische En Sociale Geografie*, 88, 1, pp. 53–64

Thomas, H., Stirling, T., Razzaque, K. and Brownill, S. (1996) Locality, Urban Governance and Contested Meanings of Place, *Area* 28, 2, pp. 186–198

Thornley, A. (1990) Thatcherism and the Erosion of the Planning System, in Montgomery, J. and Thornley, A. (eds) *Radical Planning Initiatives*, Aldershot: Gower, pp 34–48

Thornley, A. (1991) *Urban Planning Under Thatcherism*, 1st edn, London: Routledge

Tickell, A. and Peck, J. (1996) The return of the Manchester Men: men's words and men's deeds in the remaking of the local state, *Transactions of the Institute of British Geographers*, 21, 4, pp. 595–616

Tickell, A., Peck, J. and Dicken, P. (1995) The fragmented region: business, the state and economic development in North West England, in Rhodes, M. (ed) *The regions and the new Europe: patterns in core and periphery development*, Manchester: Manchester University Press, pp. 247–272.

Time and Talents (1997) *Bermondsey and Rotherhithe; A profile of a Fragmented Community*, London: Time and Talents

Town and Country Planning Association (1979) *Inner Cities*, London: TCPA

Totterdill, P. (1989) Local economic strategies as industrial policy: a critical review of British developments in the 1980s, *Economy and Society*, 18, 4, pp. 479–526

Travers, Morgan. (1972) *The Docklands Study*, London: Travers Morgan

Travers, T. (1988) Thatcher's gift to the East-Enders, *New Statesman*, 27 March, pp. 14–15

Turner, G. (1968) *The North Country*, London: Eyre and Spottiswoode

Turok, I. (1991) Policy Evaluation as Science: a Critical Assessment, *Applied Economics*, 23, pp. 1543–1550

Turok, I. (1992) Property-led urban regeneration: panacea or placebo? *Environment and Planning A*, 24, 3, pp. 361–380

Turok, I. and Wannop, U. (1990) *Targeting Urban Employment Initiatives*, Department of the Environment Inner Cities Research Programme, London: HMSO

Tweedale, G. (1995) *Steel City – entrepreneurship, strategy and technology in Sheffield*, Oxford: Clarendon Press

Tye, R. and Williams, G. (1994). Urban regeneration and centre-local government relations: the case of East Manchester, *Progress in Planning*, Oxford: Pergamon Elsevier

Tyne and Wear Chamber of Commerce (1990) *Contact*, August

Tyne and Wear County Council (1979) *Tyne and Wear Structure Plan*, Newcastle: TWCC

Tyne and Wear Development Corporation (1988) *Regeneration Statement*, Newcastle: TWDC

Tyne and Wear Development Corporation (1990) *A Vision of the Future*, Newcastle: TWDC

Tyne and Wear Development Corporation (1997) *The Power of 10*, Newcastle: TWDC

Tyne and Wear Development Corporation (1998) *Exit Strategy Document*, Newcastle: TWDC

Urry, J. (1990) *The Tourist Gaze*, London: Sage

Valler, D. (1996) 'Strategic' enabling? Cardiff City Council and local economic strategy, *Environment and Planning A*, 28, 5, pp. 835–855

Wales Millennium Centre Project (n.d) *Play your part in providing Wales and its Capital with the best Arts Facilities in the World*, Cardiff: WMCP

Walker, D. (1989), What's left to do? *Antipode*, 21, 2, pp. 133–165

Walton, D. (1990) Cardiff Bay Development, *paper delivered to the Town and Country Planning Summer School*, Swansea

Walton, J. (1990) Theoretical methods in comparative urban politics, in Logan, J. and Swanstrom, (eds) *Beyond the City Limits*, New York, Temple University Press, pp. 243–260

Ward, K. G. (1997a) Metamorphic Manchester? Regeneration in the city, *University of Manchester School of Geography Business Elites and Urban Politics Working Paper 9*, Manchester

Ward, K. G. (1997b) Birmingham: blighted but not beaten: *University of Manchester School of Geography Business Elites and Urban Politics Working Paper 7*, Manchester: University of Manchester, School of Geography

Ward, K. G. (1997c) The politics of catch-up: regeneration in Leeds, *Business Elites and Urban Politics Working Paper 8*, Manchester: University of Manchester School of Geography

Ward, K. G. (1998) Culture clash: marketing the city in the remaking of city politics, *mimeograph*, Manchester: University of Manchester School of Geography

Ward, S., (1995) *Planning and Urban Change*, London: Paul Chapman Publishing

Watson, S. and Gibson, K. (eds) (1996) *Postmodern Cities*, Oxford: Blackwell

Welsh Office (1997) *Welsh Capital Challenge and Strategic Development Scheme Guidance 1998–99*, Circular 48/97, Cardiff: Welsh Office

Wilkinson, S. (1992) Towards a new city? A case study of image improvement initiatives in Newcastle upon Tyne, in Healey, P., Davoudi, S., O'Toole, M., Tavsanoglu, S. and Usher, D. (eds) *Rebuilding the City, Property-led Urban Regeneration*. London: E&FN Spon, pp. 174–211

Wilks-Heeg, S. (1996) Urban experiments limited revisited: urban policy comes full circle, *Urban Studies*, 33, 8, pp. 1263–1297

Williams, E. (1998) *London Docklands*, Harmondsworth: Penguin

Williams, G. (1995) Manchester City Pride – a focus for the future, *Local Economy* 10, 2, pp. 124–132

Wilson, D. and Game, C. with Leach, S. and Stoker, G. (1994) *Local Government in the United Kingdom*, London: Macmillan

World Commission on Environment and Development (1987) *Our Common Future*, Oxford: University Press.

Index